Control

Leonardo

Roger F. Malina, Executive Editor

Sean Cubitt, Editor-in-Chief

See http://mitpress.mit.edu for a complete list of titles in this series.

Control

Digitality as Cultural Logic

Seb Franklin

The MIT Press
Cambridge, Massachusetts
London, England

This book was set in Stone by the MIT Press.

Library of Congress Cataloging-in-Publication Data

Franklin, Seb, 1982–
Control : digitality as cultural logic / Seb Franklin.
    p.   cm.— (Leonardo book series)
Includes bibliographical references and index.
ISBN 978-0-262-02953-7 (hardcover : alk. paper), 978-0-262-55260-8 (pb)
1. Information technology—Social aspects. 2. Cybernetics. I. Title.
HM851.F7234 2015
303.48'33—dc23
2015009273

Technology reveals the active relation of man to nature, the direct process of the production of his life, and thereby it also lays bare the process of the production of the social relations of his life, and of the mental conceptions that flow from these relations.
—Karl Marx, *Capital: A Critique of Political Economy*

[A]t the end of the twentieth century, the image of steering, that is, management, became the cardinal metaphor for describing not only politics but also all human activity.
—Tiqqun, "The Cybernetic Hypothesis"

Contents

Series Foreword

Leonardo/International Society for the Arts, Sciences, and Technology (ISAST)

Leonardo, the International Society for the Arts, Sciences, and Technology, and the affiliated French organization Association Leonardo have some very simple goals:

1. To advocate, document, and make known the work of artists, researchers, and scholars developing the new ways that the contemporary arts interact with science and technology and society.
2. To create a forum and meeting places where artists, scientists, and engineers can meet, exchange ideas, and, when appropriate, collaborate.
3. To contribute, through the interaction of the arts and sciences, to the creation of the new culture that will be needed to transition to a sustainable planetary society

When the journal *Leonardo* was started some forty-five years ago, these creative disciplines existed in segregated institutional and social networks, a situation dramatized at that time by the "Two Cultures" debates initiated by C. P. Snow. Today we live in a different time of cross-disciplinary ferment, collaboration, and intellectual confrontation enabled by new hybrid organizations, new funding sponsors, and the shared tools of computers and the Internet. Above all, new generations of artist-researchers and researcher-artists are now at work individually and collaboratively bridging the art, science, and technology disciplines. For some of the hard problems in our society, we have no choice but to find new ways to couple the arts and sciences. Perhaps in our lifetime we will see the emergence of "new Leonardos," hybrid creative individuals or teams that will not only develop a meaningful art for our times but also drive new agendas in science and stimulate technological innovation that addresses today's human needs.

For more information on the activities of the Leonardo organizations and networks, please visit our websites at http://www.leonardo.info/ and http://www.olats.org.

Roger F. Malina
Executive Editor, Leonardo Publications

ISAST Governing Board of Directors: Nina Czegledy, Greg Harper, Marc Hebert (Chair), Gordon Knox, Roger Malina, Tami Spector, Darlene Tong

Acknowledgments

This book was shaped by conversations with many colleagues and friends, several of whom read and commented on the manuscript. I am grateful to all of them, and in particular to Jane Elliott, Alex Galloway, Joss Hands, Michael Lawrence, Jussi Parikka, John David Rhodes, Steven Shaviro, and Arabella Stanger. At the University of Sussex, Peter Boxall supervised the PhD project that preceded and in many ways laid the groundwork for this book. The time I have spent at Anglia Ruskin University, the University of Surrey, and King's College London has been enriched by numerous exchanges with colleagues, especially Tina Kendall, David Skinner, Helen Hughes, Paul Gilroy, Kélina Gotman, Pat Palmer, Clare Pettitt, and Mark Turner. My heads of department at King's, Jo McDonagh and Richard Kirkland, provided exemplary support while I was working on the final manuscript. I also thank my students, especially those on the MA in Contemporary Literature, Culture and Theory at King's, for inspiring me with their commitment and intellectual curiosity.

I was able to present and discuss several parts of the book thanks to kind invitations from the Centre for Critical Theory at the University of Nottingham, the World Picture conferences at the Universities of Toronto and Sussex, the Department of Media and Film research seminar at Sussex, the London Beckett Seminar at Birkbeck, University of London, and the Past Imperfect seminar at University College London. I am grateful to everybody involved in these events for their hospitality, and for helping me to refine and reformulate much of the material that appears in the book.

At the MIT Press, Sean Cubitt and Doug Sery supported the project from proposal to publication. Susan Buckley, Virginia Crossman, and Katie Helke Dokshina provided invaluable editorial assistance and expertise at various points in the process. Four anonymous readers reviewed the manuscript and provided detailed and thoughtful comments, all of which helped me to make significant improvements to the book. I thank all of them for making the publication of this book possible.

Introduction: The Computer as Metaphor

The postindustrial society. The information economy. The third wave. Late capitalism. Post-Fordism. The network society. Neoliberalism. The new spirit of capitalism. Empire. The desire to account for the present socioeconomic moment has led to what can only be described as a frenzy of periodization. And this frenzy of periodization can only be understood as a symptom of an episteme in which the diffusion of exploited labor across the social fabric in the overdeveloped world and ever-growing rates of exploitation, expulsion, incarceration, and destruction in the fissures and at the margins of this world exist as features of a sociocultural-economic system in which the supposedly frictionless movement of information functions as a sovereign concept.[1] Each of the periodization theories listed at the beginning of this introduction grapples with one or more components of a historical arc that begins to coalesce after World War II, intensifies through the social movements of the 1960s and the currency and oil crises of the 1970s, and continues to unfold today. This historical trajectory can be analyzed along a number of different vectors: from industrial dynamics to the specific, technologically mediated practices undergirding the most recent transformations in modes of production; from new and emerging types of commodity to newly flexible and dynamic organizational diagrams; from defined political programs implemented through state legislation to the decline of various forms of sovereignty (even if, as described in Michael Hardt and Antonio Negri's book *Empire*, these return under a new, distributed, and swarmlike form).

But what kinds of historical and representational forces are at work behind this desire for periodization? What might emerge if one looks through all of these efforts to distinguish the current historical moment from its preceding era—broadly put, the period of industrial modernity, factory-line production, disciplinary institutions (the school, the hospital, the prison), national sovereignty, and so on—in order to locate the conditions of knowledge that ground them? The accounts contained in *The*

Post-industrial Society, The Rise of the Network Society, and *The New Spirit of Capitalism,*[2] for example, are thorough enough when it comes to describing the characteristics of certain practices of production and distribution, making neat and tidy distinctions between distinctive technological paradigms, observing new organizations of labor time, or tracing the changing dynamics of family life (at least, they are thorough enough when it comes to the Global North). But achieving this rigorous definition of the present and its distinction from the past comes at a cost. Asserting the radical difference between present and past without examining the contingency of the conceptual frameworks, spatial diagrams, and metaphors one uses in order to do so risks obscuring those shifts in the conditions of knowledge that are required for diffuse groups of individuals, institutions, and systems to desire, conceptualize, and enact such differences in the crucible of history.

Returning to the periodizing concepts listed earlier, one might note that several of them are grounded in specific technological substrates—automation and self-regulating machines in general, and electronic computers more specifically. It is tempting, then, to begin this inquiry from the specific forms and uses of computing machines. But if these substrates can be briefly held to one side, certain recurring motifs and structures can be seen to constitute a deep-rooted cultural logic with widespread implications for concepts as fundamental (and as disputed) as identity, thought, and the social. In other words, each of the major periodization theories centered on the late-twentieth and early twenty-first centuries nods more or less extensively to the electronic digital computer and its associated practices, but they are not histories of technology, nor are they software guides or coding manuals or guides to the effective integration of computer systems within given institutions. These accounts are concerned not with computer technologies themselves but rather with the constellation of socioeconomic transformations that surround the emergence and present ubiquity of the electronic digital computer. The questions raised by this perspective—about how thought and practice relating to the management of society have become imbricated with but not simply determined by particular technologies (because if the latter were true, such wrangling over specific features of the present period would not be necessary; one could simply itemize the ways in which computers are used in factories, offices, trading floors, schools, prisons, and so on)—are central to locating a cultural logic of the so-called information age.

Control: Digitality as Cultural Logic addresses the emergence and normalization of the conditions of knowledge that (1) make concepts such as "the information economy" thinkable and (2) determine the deleterious effects of these concepts when they are turned loose on material social spaces, from factories and offices in the overdeveloped world to the vast spaces of dispossession that both undergird the social conditions of these overdeveloped spaces and persist in their interstices. Because these conditions emerge in complex, uneven ways, and because they are produced socially and culturally as well as technically, they cannot necessarily be grasped as a totality or tracked in a linear fashion. The book thus seeks to locate their emergence across three intertwined threads. The first consists of the roots of a relationship between information, labor, and social management that emerged in nineteenth-century political economy, technology, and governmentality and became established as norms in the second half of the twentieth century; the second consists of the development and diffusion of human–computer metaphors in the middle decades of the twentieth century; and the third observes the breadth and penetration of these informatic principles in certain socioeconomic and cultural practices in the late-twentieth and early twenty-first centuries.

What kinds of assumptions are required to understand people and their multiple, heterogeneous social interactions in terms of digital information and its processing and transmission? What historical processes would be necessary to operationalize these assumptions at the level of social and political orthodoxy? And what would be the socioeconomic and cultural implications of such a vision of the world functioning as an unmarked norm? These are the questions that guide the following pages. Setting out to answer them requires a method that draws on critical theory, media theory, and the history of science. It also requires an engagement with the possibility that many of the forms of violence that exist under the present arrangement of global political economy are not accidents or problems simply waiting to be solved under the newer, more flexible, communicative, and connected economic mode, but rather features that are internal to the same logic that makes ideas of society as a communication network or an information-processing system possible in the first place.

The logic under which social worlds are reconceptualized as information-processing systems is here defined as *control*. Control, as it is theorized in this book, describes a set of technical principles having to do with self-regulation, distribution, and statistical forecasting that is extended to the conceptualization of sociality through a series of subtle historical transformations. As such, it also describes the episteme grounding late capitalism, a worldview that persists beyond any specific device or set of practices. Control, in this sense, must be understood as fundamentally digital but not necessarily confined to social practices that are directly mediated by electronic digital computers. A prefatory note is thus required to demonstrate the ways in which the concept of control pursued here encompasses but also departs from a more conventional version of the concept that has to do with technical principles of information processing and their deployment in industrial production, the emergence of knowledge work, biology, and materialist psychology, among other fields.

The existing, materialist concept of control from which this book departs describes a set of principles concerning self-regulation in animals and machines as well as a paradigm for technical and social organization gaining prominence in the overdeveloped countries after World War II.[3] Although interest in this notion of control can be seen to bloom after the publication of Norbert Wiener's book *Cybernetics: Control and Communication in the Animal and the Machine* in1948, the principle of technological self-regulation must be situated in a broader history. This history encompasses at least the use of nondigital systems such as float regulators in ancient Greece, James Watt's flyball governor for steam engines of 1788, and James Clark Maxwell's landmark paper "On Governors" of 1868. The words used to name each of these technologies must be seen as conveying more-or-less explicit political valences—*control* itself derives from the French term *contre-rolle* (counter-roll), a copy of a legal document used to verify the authenticity of the "original," and the implications of the word *governor* scarcely require elaboration. In analyses of control as material self-regulation, however, this political valence is generally rendered as coincidental to the function of particular apparatuses and practices.[4]

Although this historical context must be accounted for, it is clear that much of the conceptual and explanatory power of control in the twentieth and twenty-first centuries stems from its interconnection with the electronic digital computer. Because of this interconnection, studies of control as self-regulation tend to encompass technical procedures of industrial automation, informatic capture, distributed command, and technical management—modes of intensified efficiency and regulation of production that characterize the socioeconomic period described in so many ways in the texts listed at the start of this introduction. This understanding of control as a primarily technosocial principle undergirds historical studies such as James R. Beniger's commanding work *The Control Revolution*. Here Beniger sets out a periodization theory that situates technological control as the source of a third industrial age consisting of

a complex of rapid changes in the technological and economic arrangements by which information is collected, stored, processed, and communicated, and through which formal or programmed decisions might affect social control. From its origins in the last decades of the nineteenth century, the Control Revolution has continued unabated, and recently it has been accelerated by the development of microprocessing technologies. In terms of the magnitude and pervasiveness of its impact upon society, intellectual and cultural no less than material, the Control Revolution already appears to be as important to the history of this century as the Industrial Revolution was to the last.[5]

Beniger's account of control as a set of technical processes that directly affect social and economic transformations is rich and compelling, covering developments in biology, psychology, and linguistics as well as technical practices in industrial production and the collection, storage, and processing of information. This aspect of control is central to the analysis of cultural logic given here. But the sense in which the term *control* is used in this book departs in important ways from those analyses that center on information technologies and their direct use or even on the wider (but still more or less direct) sociopolitical implications of these technologies and their uses. In short, this book is as concerned with control's fuzzier sociocultural valences as it is with the technical and organizational functions from which these valences are derived. These sociocultural aspects operate in concert with control's surface technical and organizational elements, and they both shape and obscure many of its more troubling effects.

At this point, it must be acknowledged that a sociocultural valance of control *can* be observed in Beniger, but in the form of an assumption that undergirds his method rather than as an object of study in itself. For Beniger, the importance of the control revolution stems from its fundamental commensurability with the biological and social structure of humanity:

Because societies [like living organisms] must also be concrete open systems if they are to sustain their organization against the progressive degrading of their collective energy, the view of organisms as concrete open processing systems applies equally to their social aggregates. The essence of human society, in other words, is its continuous processing of physical throughputs, from their input to the concrete social system to their final consumption and output as waste.... Unlike living organisms, however, *social* systems are made up of relatively autonomous components—individuals, families, groups, organizations—that can act for different and even cross-purposes. Because system processing must depend on exchanges among these individual components, the need for their coordination and control means that information processing and communication will account for a greater proportion of matter and energy flow than they do in single organisms. The actual proportion will depend on several factors, including size of the population and its spatial dispersion, complexity of organization, and volume and speed of processing, among others.[6]

Beniger's application of principles derived from the electronic digital computer to fundamental conceptualizations of individual human actors and
social groups is symptomatic of a historical shift that the political scientist Karl Deutsch described in 1963 as moving the "center of interest from
drives to steering, and from instincts to systems of decisions, regulation,
and control."[7] The wider logic of control that grounds informatic capitalism, then, reaches beyond the direct social and technological practices
whose growth, beginning in the late nineteenth century, is accounted for
by writers such as Beniger, Wiener, and David Mindell. The logic of control
as episteme describes a wholesale reconceptualization of the human and of
social interaction under the assumption—visible in Beniger's work from the
1980s as well as in the dominant social, economic, and political practices
of the present—that information storage, processing, and transmission (as
well as associated concepts such as "steering" and "programming") not
only constitute the fundamental processes of biological and social life but
can be instrumentalized to both model and direct the functional entirety
of such forms of life.

Where for writers such as Beniger (and for economists such as Gary Becker
and Friedrich Hayek) this principle of control serves as an objective ground
from which further analysis can be conducted, in this book it emerges as an
object of contestation. It describes a deleterious process though which, as
Tiqqun put it, steering (a term that can be taken as a synonym for control
in all of its technical, political, and cultural valences) becomes the *guiding
metaphor for all human activity*.[8] Control, this book argues, should be understood as the logical basis of a worldview that imbricates literal practices
of computation, the new organizational and infrastructural concepts these
practices facilitate, and metaphors derived from the electronic digital computer and its processes with a system of value production that can produce
profit only by exploiting and dispossessing human life.

For this epistemic movement to become thinkable requires a turn toward
digitality not only as a logical-technical substrate through which certain
machines might operate but also as a predominant logical mode with
which to address both individual social actors and the body of interactions
between these actors that can be dubbed "society." Digitality can be placed
in the category of practices that Bernard Stiegler describes as proceeding
through grammatization, "the process through which the flows and con-
tinuities which weave our existences are discretized."[9] The digitality that
informs control, though, goes beyond the other practices of grammatization Stiegler addresses (writing, the rationalization of production through
machine tools) in that it no longer presents discretized representations as

systems that imply a more continuous or complex world behind them; instead, the logic of control posits its objects as already fundamentally discrete, at least in every way that can possibly matter. To be clear, what is being formulated here is not an argument in favor of any specific, continuous concept of life, sociality, or the physical universe, although, as N. Katherine Hayles suggests in *My Mother Was a Computer*, arguments to the contrary are certainly part of the control episteme.[10] Rather, this book is an inquiry into the ways in which certain digital conceptualizations of those phenomena emerge, are normalized, and function within social, political, and cultural practices. In this sense, the movement from digital systems to control is comparable to the movement in the concept of networks that, for Luc Boltanski and Ève Chiapello, underpins their "new spirit of capitalism." For Boltanski and Chiapello, the contemporary ubiquity of the word *network*, formerly associated with the either technical distribution of resources (water, electricity) or with secret organizations (resistance, trafficking) amounts to a "rehabilitation" that can be understood only in the context of the new working methods facilitated by computer technologies, but that cannot be reduced to these technologies and methods.[11] Digitality, as it functions under the sign of control, comes to describe not only a set of technologies or logical operations but also a fundamental condition. To exist, from the point of view of control, is to be digital—or, in Friedrich Kittler's words, after the emergence of digital signal processing, "only what can be configured as a switching circuit exists."[12] This ontological digitality, separated from the machines and interfaces with which it has become synonymous, entails a fundamental process of discretization that can be purely conceptual as much as it can enable particular technological processes.[13] In the same way, digitality moves from a specific system of representation and technical processing to a set of generalized metaphors. The historical emergence of control, as it is characterized in this book, accounts for the processes through which this generalization of digitality takes place.

It is in the context of digitality's becoming-general that the tensions raised at the outset of this chapter—between "creative" information work and the expansion of paid and unpaid labor as well as between the "immaterial," frictionless information society and the constant violence of exclusion, expulsion, and incarceration that is the corollary of this society—must be situated. Digitality promises to render the world legible, recordable, and knowable via particular numeric and linguistic constructs. For this rendering of the world to take place, however, there must be processes of capture, definition, optimization, and filtering that necessarily implement a distinction between those aspects of the world that are intended and included within a given digital representation and those that are excluded or filtered out.

The process of digitization as well as the necessary process of exclusion it entails can be illustrated through the technique of pulse code modulation (PCM). Developed at Bell Laboratories in 1943 following the discovery of a patent held by Alec H. Reeves since 1938, PCM represents the first technique for the digitization of analog signals such as speech, recorded sound, and images.[14] PCM consists of two stages. The first stage samples the analog signal, dividing it into a number of discrete units that must be at least double the signal's highest frequency. There is, then, an immediate imposition of uniform, discrete steps onto continuous matter. As B. M. Oliver, J. R. Pierce, and Claude Shannon put it, "[T]o transmit a band-limited signal of duration T, we do not need to send the entire continuous function of time. It suffices to send the finite set of $2W_oT$ independent values obtained by sampling the instantaneous amplitude of the signal at a regular rate of $2W_o$ samples per second."[15] This first stage results in "a definite and limited number of amplitudes per unit of time which replace the original wave in subsequent operations."[16] It also allows for the exclusion of "noise" by dividing the signal into "desired" bands and those that fall outside of them: as the communication engineer W. M. Goodall puts it, "When the sampling frequency is at least twice the highest frequency present in the original wave, the resulting distortion falls outside the desired band and can be removed by a low-pass filter in the output of the system."[17] The second stage of PCM quantizes the signal's amplitude by reducing its total field of possibility to a fixed set of permitted values, each expressed as a sequence of on/off states.[18] Like the sampling stage, this second stage is founded on a process of limitation and, by extension, exclusion. "By quantizing," Oliver, Pierce, and Shannon write, "we limit our "alphabet.""[19] These processes of filtering, averaging, and excluding, which are inherent to the digitization of continuous phenomena, appear innocuous enough when applied to telephone or television signals but become fatal when the computer is idealized and deployed as a model for socioeconomic management.

In "There Is No Software" Friedrich Kittler observes that the tension between the ideal principle of computation as theoretically universal and the basic procedures of excluding or filtering that material practices of computation actually entail is fundamental to digital culture, being traceable at least to Alan Turing's foundational 1936 paper on computable numbers and their applicability. Considering the gap between the symbolic register of the computer and the material fullness of the real, Kittler writes that only in Turing's paper "On Computable Numbers with an Application to the *Ents-cheidungsproblem*" does there exist a machine with unbounded resources in space and time, with infinite supply of raw paper and no constraints on computation speed. All physically feasible machines, in contrast, are limited by these parameters in their very code. The inability of Microsoft DOS to tell more than the first eight letters of a file name such as WordPerfect gives just a trivial or obsolete illustration of a problem that has provoked not only the ever-growing incompatibilities between the different generations of eight-bit, sixteen-bit and thirty-two-bit microprocessors, but also a near impossibility of digitizing the body of real numbers formerly known as nature.[20]

In other words, although a concept of the computer as universally inclusive might be theoretically possible, the notion that real (which is to say non-discrete) phenomena can be digitized without exclusion is precluded by the materiality of actual computing machines. The persistent reformulation of the computer from finite, concrete technology to universal metaphor (for the brain, for the subject, for the economy, for society) lies behind the emergent logic of control and as such is central to the historical phenomena addressed in this book.

Addressing this construction of digitality as metaphor imposes a specific set of methodological concerns. The pejorative concept "vapor theory" has been central to the development of materialist media theories, especially those pertaining to electronic digital computers.[21] Vapor theory, in these accounts, describes critical analyses of technology that are based on vague metaphors rather than on rigorous materialist engagement—or, as Peter Lunenfeld puts it, "critical discussions about technology untethered to the constraints of production."[22] Yet conceptual deployments of computer technology that are not tethered to the constraints of their materiality proliferate in the socioeconomic formulations of the past sixty years, from ideological deployments of genetics that frame market competition as natural to ideas such as that of the network society in which social actors are conceived of as interrelating like computers.

Control thus begins from the following principle, which necessitates a synthetic engagement with *both* vague metaphors *and* materialist analysis: even in the age of so-called immaterial economies, the dominant mode of production cannot be reduced to the dead labor of machinery but rather rests on a complex of technical processes and socioeconomic and cultural logics that rely on vague metaphors drawn from such technologies for their conceptual efficacy. In this arrangement, the technical processes facilitate specific working practices, while the vague metaphors allow for the constant expansion of processes of valorization as the subject is reconceptualized as both a communication system and a component in such a system. Communicative acts are recast and exploited as labor, and already existing forms of labor are recast as communication. If the vagueness of the human–machine metaphor is central to its mobilization by capital, one cannot discard it but instead must identify it (and thus describe it) and then critique it through (1) historical and material specificity and (2) the extraction of a political valence from the specific account. Put simply, vapor theory cannot be ignored when it constitutes the cultural layer of the dominant mode of production; it must be engaged and traced both to its origins and to the desires that it reveals.

This book seeks to identify control as episteme (rather than as confined to certain fields) by locating its historical grounding and its foundational logic across a diverse body of objects and practices, from cybernetics to economic theories and management styles, to concepts of language and subjectivity, to literary texts, films, and video games. This breadth is essential because when the digital is brought to bear on concepts such as labor, subjectivity, and collectivity, it constitutes not only a set of technical processes but also a social logic that reaches beyond computational technologies. This commingling of the technical and the social is here defined as foundational to the logic of control. Control both defines and instrumentalizes individual actors and groups, whose conditions of social existence are now premised on statistical predictive models and decisional states that rest on a conceptual as well as a technical digitization of the world. Following Fredric Jameson's lesson that explicit political content cannot, taken alone, divulge the true political implications of a cultural object,[23] it thus becomes necessary to look not only at but also through the specific technical objects, economic practices, industrial formations, political ideals, and organizational diagrams of the present in order to trace the digital logic of control that animates and facilitates all of these phenomena under the banner of the so-called information economy (or whichever epithet one chooses to apply to this era).

I Digitality without Computers

1 Control

To begin untangling the dense web of relations between control, digitality, and capital, we must turn to a small body of writing that has perhaps been conspicuous by its absence from the previous pages. In the late 1980s and early 1990s, Gilles Deleuze began to write of a distinctive historical period that he came to describe as that of control societies. This notion of control societies, although traceable to certain concepts espoused in *A Thousand Plateaus* and *Cinema 2*, comes into focus in the final pages of Deleuze's 1986 book on Michel Foucault and is developed through the essay "Having an Idea in Cinema" (published in 1990 but previously delivered as a lecture in 1987), a discussion with Antonio Negri published in 1990 under the title "Control and Becoming," and a short piece from the same year titled "Postscript on Control Societies."[1] Control, for Deleuze, describes the apparently "free floating" modes of organization that emerge after the institutions that Foucault examines in *Discipline and Punish*—the factory, the prison, the hospital, the school, the family—begin to break down.[2] Although Deleuze locates and itemizes many features and symptoms of control, he stops short of examining its historicity, and only briefly hints at its underlying logic. Nonetheless, his compressed elaboration of the concept both incorporates a number of features arising across several more extensive periodization theories (including those listed in the introduction to this book) and implies an epistemic grounding that many of these latter theories pass over. As such, it serves as a foundation for the expanded concept of control developed here.

Starting from the second of the justifications detailed above, it is clear that in Deleuze's theorization the epistemic character of control is grounded in the intersections of technology and knowledge raised in the introduction to this book. For all of his references to material technologies such as "computers," "information technologies," and "electronic card[s]," Deleuze is quite clear that the concept of control, although undoubtedly intertwined with these machines, accounts for a far wider set of socioeconomic logics and practices undergirding the characteristic impositions of the current stage of global capitalism. This wider meaning is underscored when Deleuze observes in "Control and Becoming" that *"cybernetic machines"* as well as computers are emblematic of control.[3] According to the logic of the interdisciplinary field of cybernetics, which provides control with a principal logical endowment and is discussed further in chapter 2, an individual human or animal, a brain, a social group or a group of such groups, a complex of interlocking markets, and a battlefield are intelligible and analyzable as self-regulating machines, just as a computer is. This is why, for Deleuze, computer technologies (like any other type of machine) cannot be seen as determining the social conditions that control underwrites.[4] Rather, such machines "express the social forms capable of producing them and making use of them."[5] Or, as Deleuze puts it in his conversation with Negri, "[T]he machines don't explain anything, you have to analyze the collective apparatuses of which the machines are just one component."[6]

These machines, though, are clearly implicated in the Deleuzian concept of control, and it is understandable that a number of theorists have sought to develop Deleuze's periodization along technological lines. Michael Hardt and Antonio Negri discuss societies of control in general technological terms, albeit extended to new forms of labor and political organization, when they write of power that is "now exercised through machines that directly organize the brains (in communication systems, information networks, etc.) and bodies (in welfare systems, monitored activities, etc)."[7] Taking a lead from Deleuze's assertion that computers are the emblematic technologies of control, Alexander R. Galloway's book *Protocol: How Control Exists after Decentralization* signals a move toward a more materially rigorous engagement with the mechanisms of control, aiming to "flesh out the specificity of this third historical wave by focusing on the controlling computer technologies native to it," specifically the technical substrate of the Internet and the relationship between bioinformatics, artificial intelligence, and biopolitics.[8] In *Control and Freedom: Power and Paranoia in the Age of Fiber Optics*, Wendy Hui Kyong Chun engages in a sympathetic critique of Deleuze, describing his analysis as "arguably paranoid" because of the way in which it appears to overestimate the technical potential of computation as a mode of social regulation and thus unintentionally fulfills the "aims" of control.[9] These accounts of control as a set of technically mediated relations are important for the work they do to flesh out Deleuze's skeletal accounts, and this book is in part built on their methods and insights. Control, though, persists as a logic beyond and behind computer technologies and the new types of labor they facilitate—a fact that should be clear from

Deleuze's location of Franz Kafka's novel *The Trial* (1925) as a canary registering cultural traces of the emerging episteme—and it is the historical and cultural character of this logic that must be addressed if its power and pervasiveness are to be properly apprehended.

Deleuze's sketch of control as a socioeconomic logic that sits behind and beyond computer technologies is fascinating for a number of reasons, not least the extraordinary range of historically specific characteristics accounted for, necessarily in passing, in such a small number of pages. Across the writings on control societies, it is possible to find the core principles of many of the grand periodization theories developed to describe the socioeconomic character of the past fifty years or so. "Speech and communication" in control societies, as in information economies with their knowledge workers, "are thoroughly permeated with money—and not by accident but by their very nature," writes Deleuze.[10] He describes the organizational diagram of control as a "*modulation* ... like a self-transmuting molding constantly changing from one moment to the next, or like a sieve whose mesh varies from one point to another."[11] What is this "modulation" if not the distributed-network form that underpins Manuel Castells's network society with its space of flows or Boltanski and Chiapello's "reticular" new spirit of capitalism?[12] In control societies, Deleuze writes, industrial production is displaced to the underdeveloped world and replaced in the overdeveloped nations by a "metaproduction" that "no longer buys raw materials and no longer sells finished products" but instead sells "services" and buys "activities"—a set of developments that, even it its extreme generality, evokes the signatures of post-Fordist economism.[13] In surveying the labor dynamics of control, Deleuze finds wages in "a state of constant metastability punctuated by ludicrous challenges, competitions, and seminars" as well as the imposition of "an inexorable rivalry presented as healthy competition" across bodies of workers.[14] Here, as well as in accounts of debt as a principle mode of confinement and of "'business' being brought into education at every level,"[15] it is not difficult to identify the conditions of universal competition that are foundational to what has been defined as neoliberalism.

These core features of control, which are socioeconomic and cultural as well as technological (as if there could be such a thing as a purely technological sphere separate from society and culture), make it clear enough that Deleuze's real focus is on the persistence and mutability of capitalism. The sly references to "moles" that are replaced by "snakes" in the control era as well as references to factories and businesses should make this historical-materialist dimension of control clear enough even before one reaches the passages in "Postscript" in which Deleuze speaks of capitalism by name.[16]

The mutation Deleuze accounts for with the term *control* can thus be summarized as a diffusion or atomization of the logic of capital, a process that is evidenced, for Deleuze, in the dispersal of the factory system into the "soul" or "gas" of the business network and the augmentation of value generated through commodity-producing labor with the codes and fluctuations of finance capital, informatics, and knowledge economies.

It is this foundational connection to capital that determines the true stakes of any inquiry into control. For all of the technological and organizational details the "Postscript" passes over, the most pressing rejoinder in it comes when Deleuze moves beyond the specificities of contemporary production to remind his readers of the "three quarters of humanity in extreme poverty, too poor to have debts and too numerous to be confined," and of the "mushrooming shantytowns and ghettoes" that remain, as ever, the direct outcome of global capitalism, be it nominally industrial, postindustrial, or what have you.[17] This is why continued inquiry into the logic of control remains essential: not only as an inventory of the newest, glossiest modes of technological organization and reproduction and their ideological framing but also as an index of the ways in which these modes, for all of their purported immateriality, flexibility, and freedom, remain built on exploitation and the degradation of material environments on a global scale. In short, any inquiry into the formal and aesthetic concerns specific to the control era can be important only insofar as it functions as a way of grasping the historical conditions that underpin the continued destruction of life and matter under the mutating logic of capital.

Digitality

In examining the relationship between technical medium, political economy, and social formation, one might appear to be on familiar territory, facing a critical practice that entails simply extending to computer culture Jean-Louis Comolli's pronouncement that the appearance and disappearance of cinematic techniques over time depend "not on a rational-linear order of technological perfectibility nor an autonomous instance of scientific 'progress,' but much rather on the offsettings, adjustments, arrangements carried out by a social configuration in order to represent itself, that is, at once to grasp itself, identify itself and itself produce itself in its representation."[18] Where Comolli's focus is on cinema and classically Althusserian notions of ideology and ideology critique, however, the complex of computation and control foregrounds a general movement toward concepts of automation—not only the automation facilitated by literal machines but also the historical emergence of the correlated idea that thought, identity, social relations, economics, and production can be conceived of as automated or self-regulating processes—that stage a mock triumph of symbolic logic, promising to do away altogether with such partisan human affairs as ideology (and, by extension, political and economic exploitation) and thus appearing to do away with the need for critique or demystification.[19] Rather than looking behind the object form of devices and practices, according to this logic, one only has to count the data and retrace the steps of the algorithm. This assurance is, of course, an illusory promise grounded in the very digital logic that undergirds both the technical and the sociocultural organization facilitated by control, from computer hardware and software to the conceptualization of thought, life, and social interactions as inherently digital and valorizable. The historical shift toward this logical framing is behind one of the principal tensions arising out of control: the tension between the promise that unfreedom, injustice, and inequality have been done away with by a (nonpartisan, multilateral, rational, and machinic) digital logic that renders all affairs and objects equal, on the one hand, and the ways in which this same logic simply increases the granularity, efficacy, subtlety, and rigor with which exploitation and exclusion are affected, on the other. It is through tracing both the historical emergence and the contemporary cultural manifestation of this digital social logic that the work of critique in the era of control must proceed.

It is clear enough that the emergence to ubiquity of electronic digital computers is not exactly a source of control societies, but rather one of several technological, social, and economic threads that are at base "deeply rooted in a mutation of capitalism."[20] And it is also clear enough that this mutation of capitalism—its radical expansion in space and time—is exemplary of what Marx calls "real subsumption," the process by which social relations formerly outside of the accumulative regime of capital are brought into this regime. It is this persistence through mutation of capital that is at stake in the recent frenzy of periodization. Whether seeking ways to get ahead of the wave and to profit or looking for ways to diagnose and/or arrest the newest forms of exploitation, any attempt to grapple with the socioeconomic transformations of the late twentieth and early twenty-first centuries is at base an attempt to understand the current episode in "that uninterrupted narrative" of capitalism that Jameson identifies as constituting the political unconscious of all cultural production.[21] What is absent from these accounts and what is necessary in order to come to terms with the linked political and cultural implications of control is an engagement with the specific relationship between digitality and the "mutation of capitalism" through the real subsumption of thought, communication, identity, aesthetics, and so on—up to and including life itself.[22]

The definition of control as epistemic grounding that functions over and above specific instances and uses of computer technologies brings a concept of the digital as logical substrate to the fore. In "Postscript," Deleuze states that the "language" of control is digital, in contrast to the analog language of disciplinary institutions. It would, then, be easy enough to proceed from this definition to the principle that digital machines directly determine the cultural implications of control—and, by extension, that it would be possible to grasp these implications simply be closely analyzing the forms and functions of digital machines. The discussion in the preceding paragraphs should make it clear that such an analysis will not be sufficient. Control is not limited to those moments when specific (computer) technologies are in use, even if these technologies are understood as historically imbricated with the social and economic practices that drive this framing of the world. Were this the case, the pervasiveness and reach of control would be severely limited and could not possibly constitute the severest sociopolitical implications that Deleuze describes, under which "we might come to see the harshest confinement as part of a wonderful happy past."[23] Rather, the digital, in Deleuze's account of control, describes a logic that comes to define fundamental elements of social existence in a range of categories, leading, for example, to the "dispersal" of certain notions of "labor," "language," and "life" that previously characterized the disciplinary era.[24]

Digitality, as evoked in the introduction to this book, can in part be understood as a mode of capturing individual and social behaviors for the purpose of valorization. Stiegler appears to support such a definition, writing of an equivalence between the "grammatization of gesture" and "proletarianization" as well as of the rearrangement of *"psychic organization"* through the "intermediary of mnemotechnical organs, amongst which must be included machine tools … and all automata" that facilitate this "proletarianizaton."[25] The finer points of both the digital mode of representation and the theoretical responses to it are addressed in greater depth in the following chapters, as are many of the specific practices underpinning the economic logic of post-Fordist valorization. At this point, it is necessary to grapple with the conundrum of how social formations are both prepared for and shaped by the logic of discretization. In other words, to engage more fully with the "mutation of capitalism" that Deleuze addresses in "Postscript," one must engage at once with this mutation's technics, politics, and (to grasp the fullness of its absorption into the social fabric of the present) aesthetics.

In the narrative set out in "Postscript," factories in the nineteenth and twentieth centuries were analogical not just because they were premised on such thermodynamic concepts as energy and the individual worker's ability to perform their labor in the form of continuous movements. Nor were factories (and, by extension, the disciplinary societies within which they present the central site of production) analogical because of the linearity of their processes (in fact, it would be more convincing to argue that the Taylorist principle of scientific management and the Fordist division of labor are fundamentally digital processes—as discussed in subsequent sections on Marx and Babbage). Rather, the factory is analogical *in the social relations it creates*—it forms "individuals into a body of men for the joint convenience of a management that could monitor each component in this mass, and trade unions that could mobilize mass resistance."[26] The processes through which labor is captured, broken down into discrete actions and units of time, and valorized might be understood, then, to exist in tension with this analogical sociality. The modes of production central to control, by contrast, are digital through and through. In control societies, the "body" of workers is subject to ever-deepening processes of discretization, first at the level of the group both inside and outside the workplace and then at the level of the individual. This is first hinted at in a game-theoretical individuation of all-against-all competition characteristic of neoliberal economism; where the factory consists of "a body of men whose internal forces reached an equilibrium between the highest possible production and the lowest possible wages," the business constantly introduces "an inexorable rivalry presented as healthy competition."[27]

But this individuation through competition does not by itself account for the socioeconomic logic of control. Rather, such processes of individualization, incentivization, and competition—which have been described in several quarters as "neoliberalism"—account for only a part of the world picture of control. Beyond the atomization of the workforce (if not of the social body more generally) into discrete units, or sovereign individuals, Deleuze comments on the formation of the *dividual* that occurs when control apprehends the social actor, "dividing each within himself."[28] The dividual is what might formerly have been understood as "the subject" once it has been divided within itself, broken down into discrete parts that are each representable as symbolic tokens and capturable as labor—a process that might begin with the rationalization of work along Taylorist principles but which intensifies to include productivity, appearance, genetic traits, lifestyle preferences, and cultural and creative faculties. In short, the dividual is the subject *digitized*. And control is the episteme of the dividual.

The question of what becomes of those elements of identity and social experience that are filtered out when this digitization takes place undergirds many of the central contradictions of capital in the information age.

In the most obvious sense, the principles of equivalence and exchangeability that make digital information so powerful as a regime of knowledge and action are also the fundamental principles of capital. As Jameson points out in the opening chapter of *Representing Capital*, the fundamental riddle masked by the social relations of capital and discussed in part I of *Capital* is that of how two qualitatively different things can be framed as quantitative equivalents of each other.[29] The theorization set out by Marx in response to this riddle is at the heart of the entire critical project of *Capital*. Marx identifies the basic digital premises of discretization and equivalence in a famous passage in which he discusses the exchangeability of distinct items based on the quantity of abstract labor time congealed in them:

If then we disregard the use value of commodities [as the source of their relative value], only one property remains, that of being properties of labour. But even the product of labour has already been transformed in our hands. If we make abstraction from the use-value, we abstract also from the material constituents and forms which make it a use-value. It is no longer a table, a house, a piece of yarn or any other useful thing. All its sensuous characteristics are extinguished. Nor is it any longer the product of the labour of the joiner, the mason, or the spinner, or of any other particular kind of productive labour. With the disappearance of the useful character of the products of labour, the useful character of the kinds of labour embodies in them also disappears; this in turn entails the disappearance of the different concrete forms of labour. They can no longer be distinguished, but are all together reduced to the same kind of labour, human labour in the abstract.[30]

This movement from quality to quantity is the fundamental procedure of abstraction that undergirds commodity formation, which is to say it is the fundamental procedure that undergirds the exploitative logic of capitalism: as Marx goes on to note, the crystallized forms of this abstract labor are "commodity-values [*Warenwerte*]."[31]

If this crystallization of time is generalized beyond the concrete productive arrangements he was able to observe in the middle of the nineteenth century—for example, the kinds of factory system that facilitate only a partial version of what is potentially intelligible as a total digitization of lived time— Marx's account of abstraction appears remarkably similar to the account of discrete numerical representation as a cornerstone of "new media" found in Lev Manovich's *The Language of New Media*. Here Manovich argues that the modularity, automation, variability, and transcodability of new media are all built on the fact that, as a first principle, digital objects are represented on a single, uniform scale of discrete values. And when the new media object originates outside the computer, just as when human activity originated outside of capitalism, it must be subject to a process of conversion:

> Converting continuous data into a numerical representation is called *digitization*. Digitization consists from two steps: sampling and quantization. First, data is [*sic*] *sampled*, most often at regular intervals, such as the grid of pixels used to represent a digital image. Technically, a sample is defined as "a measurement made at a particular instant in space and time, according to a specified procedure." The frequency of sampling is referred to as *resolution*. Sampling turns continuous data into *discrete* data. This is [*sic*] data occurring in distinct units: people, pages of a book, pixels. Second, each sample is *quantified*, i.e. assigned a numerical vale drawn from a defined range (such as 0–255 in the case of an 8-bit greyscale image).[32]

What is this process if not a restaging of the conversion of human time into individual units on a single, abstract scale, which can then be used to work out an equivalence between qualitatively different objects (wheat, steel, etc.), but at a more universal level? This dream of universal digitalization and valorization is the dream of capital, and the logical system through which it proceeds is that of control.

In his introduction to the 1976 *New Left Review* edition of Marx's *Capital*, Ernest Mandel makes a compelling observation, writing that "from a structural point of view, the 'concrete' capitalism of the final quarter of the twentieth century is much closer to the 'abstract' model of *Capital* than was the 'concrete' capitalism of 1867."[33] The control stage of capitalism that produces concepts such as "informational," "immaterial," "cognitive," "affective," and "networked" labor, Mandel here suggests, might not be a mutation at all. It may in fact represent a fuller realization of the abstract logic of capital than the actually existing processes of production from which Marx discerned this logic. If, as Marx writes in the *Grundrisse*, the laborer from the point of view of capital "does not represent a condition of production, but only labour" that could just as well be performed by machinery or even by water or by air,"[34] then it follows that a full realization of this viewpoint would posit all activity as labor regardless of where or when or how it takes place. Following this, a relationship between digitality, the abstract logic of capital, and the concrete practices through which this logic is enacted might be historicized by mapping the two-stage procedure of digitization that Manovich identifies (as do many others, not least Oliver, Pierce, and Shannon) onto the conditions of industrial production about which Marx writes.

Put simply, this two-stage process of *sampling* and *quantization* might be understood in terms of rationalized labor time and the compression of the fullness of lived experience according to the conditions of imposing this time. Commodity production, Marx writes in the *Economic Manuscript of 1861–63*, would be unimaginable without the uniform, discretized time made possible by the clock, from which comes "the idea of the automatic mechanism and of automatic motion applied to production." Without the clock, Marx asks, "what would be a period in which the value of the commodity, and therefore the labour time necessary for its production, are the decisive factor?"[35] This formalization of work according to a mechanically impressed temporality is, for Foucault, the central objective of disciplinary practices, which must "be understood as machinery for adding up and capitalizing time."[36] In short, it is this mechanically impressed time that allows for the quantification of human life as labor, as Marx writes in *The Poverty of Philosophy*:

If the mere quantity of labor functions as a measure of value regardless of quality, it presupposes that simple labor has become the pivot of industry. It presupposes that labor has been equalised by the subordination of man to the machine or by the extreme division of labor; that men are effaced by their labor; that the pendulum of the clock has become as accurate a measure of the relative activity of two workers as it is of the speed of two locomotives. Therefore, we should not say that one man's hour is worth another man's hour, but rather that one man during an hour is worth just as much as another man during an hour. Time is everything, man is nothing; he is, at the most, time's carcase [*sic*]. Quality no longer matters. Quantity alone decides everything; hour for hour, day for day; but this equalizing of labor is not by any means the work of M. Proudhon's eternal justice; it is purely and simply a fact of modern industry.[37]

What distinguishes the "concrete" conditions of nineteenth-century production from a theoretically full deployment of capital's abstract logic is the fact that under the former the process of digitizing life as labor is thinkable only in particular times and spaces—most obviously the bounded time of the working day and the confined space of the workplace in which rationalization can be affected mechanically. This bounded time–space of production implies a presupposition that the temporality of human life is not inherently discrete and that some device or system is required to represent it as such (to digitize it) and to thus render it subject to congealing into commodity form.

The epistemic tension between life and labor is foregrounded in Marx's work by the concept of alienation. In defining the constituent features of alienated labor, Marx writes that "labor is *external* to the worker, i.e., it does not belong to his intrinsic nature," and that because of this the worker "does not affirm himself but denies himself, does not feel content but unhappy, does not develop freely his physical and mental energy but mortifies his body and ruins his mind." This alienation leads to a formal division between labor and leisure, between the space in which labor takes pace and all other spaces, because the worker "only feels himself outside his work, and in his work feels outside himself."[38] This division, Marx suggests, presents a limit to valorization, or to the spaces and times in which the worker can be apprehended as labor by capital: because labor is "not the satisfaction of a need" but "merely a *means* to satisfy needs external to it," the alienated worker shuns work "like the plague" as soon as no "physical or other compulsion exists."[39] Life cannot be posited as labor across the full spectrum of social interactions because the worker's activity "operates on the individual independently of him—that is, operates as an alien, divine or diabolical activity" and as such functions as a "loss of self."[40] György Lukács directly connects this loss of self to the rationalization—which is to say, the digitization—of life in the workplace when he writes:

In this environment where time is transformed into abstract, exactly measurable, physical space, an environment at once the cause and effect of the scientifically and mechanically fragmented and specialised production of the object of labour, the subjects of labour must likewise be rationally fragmented. On the one hand, the objectification of their labour–power into something opposed to their total personality (a process already accomplished with the sale of that labour–power as a commodity) is now made into the permanent ineluctable reality of their daily life. Here, too, the personality can do no more than look on helplessly while its own existence is reduced to an isolated particle and fed into an alien system. On the other hand, the mechanical disintegration of the process of production into its components also destroys those bonds that had bound individuals to a community in the days when production was still "organic." In this respect, too, mechanisation makes of them isolated abstract atoms whose work no longer brings them together directly and organically; it becomes mediated to an increasing extent exclusively by the abstract laws of the mechanism which imprisons them.[41]

The clock divides time into discrete units, and the workplace and its machines compress life into functions of production. Everything outside of this defined time–space of work is conditioned by the modes of social organization that support it but nonetheless escapes procedures of direct valorization (although, as Silvia Federici, Leopoldina Fortunati, and others have shown, this escape does not automatically connote an escape from exploitation or from the many deleterious conditions that arise from implication in the overall production-reproduction cycle).

The question of how processes of digitization–valorization apprehend life beyond the nominal workplace can be addressed from a few directions.

It can, for example, be productively illuminated through a consideration of the concept of *capture*. Capture, as an instrumental practice in the digital rendering of the world, accounts for a basic form of relation between capital and the broader category of digital representation in the present socioeconomic moment. Deleuze and Félix Guattari extensively discuss the historical relationship between the capture of movement and space, on the one hand, and the logic of capital, on the other, in the chapter "Apparatus of Capture" in *A Thousand Plateaus*, and Giorgio Agamben provides a broad definition of the Foucauldian term *apparatus* as "literally anything able to capture, orient, determine, intercept, model, control, or secure the gestures, behaviors, opinions, or discourses of living beings."[42] Information work can thus be understood as proceeding through processes of *informatic capture*. In "Surveillance and Capture: Two Models of Privacy," Philip E. Agre analyzes the range of processes through which labor activities in the late twentieth-century organization are not only tracked by computer systems but also fundamentally restructured to better fit with these systems' representational and functional schemas. Capture, for Agre, describes the imposition of a computational ontology onto the world of human activity and material space: it is principally manifested in the "practices of information technologists," is "built upon linguistic metaphors," and "takes as its prototype the deliberate reorganization of industrial work activities to allow computers to track them in real time."[43]

Agre's later claim that "when a technical community with a definite worldview, perhaps puts that worldview into practice … any given metaphor will necessarily consign certain aspects of reality to the practitioner's peripheral vision" thus emerges as a materialist valence of the concept of control as episteme.[44] In making this claim, Agre is clearly referring to information technologists such as software engineers and interface designers. But if one expands the notion of a "technical community" to describe the complex of human and nonhuman actors that constitute and dispense the digital worldview of late capitalism, then the question of what is central (and thus captured and modeled) and what is peripheral (and thus discarded) within computationalist modes of social representation takes on a distinctive historical and political significance.

Agre's work is compelling for two reasons: it centers the digital logic of capture not on computers but on human behavior (wherein the role of computers for tracking provides only a "prototype" for ideal behavior rather than a final application); and it suggests that the mutation of capitalism that produces the information economy lies not only in the direct use of specific new technologies, but also in the deployment of the linguistic metaphors employed by information technologists as a set of principles through which social activity *in general* can be rendered. In short, by closing the conceptual gap between the social actor and the computing machine, the logic of control (here focused through the concept of capture) dissolves the tension between life and labor that limits the extraction of value from the former though the latter. In Agre's account, this conceptualization of capture as a reorganization of life according to the principles of computation is dispensed mainly through algorithmic procedures and hence appears as an extension of Taylorist logic into practices of navigating space and time between fixed sites. The abstract principle of conceptualizing the world digitally in order to extract value from it can, however, be seen extending beyond obvious labor practices such as computer-facilitated factory production and shipping and logistics (to draw on just two examples of the process given by Agre) across a number of political-economic accounts that emerge against the backdrop of control.

At the level of economic theory, the capture and valorization of various forms of human activity beyond the nominal time and place of work, as well as the corollary extension of economic mechanisms to function as a general representational scheme, can be observed as the logical backbone of what has been called "neoliberalism." Gary Becker's work on the extension of economic theory to social behavior, crystallized in the theory of human capital, is exemplary in demonstrating both the totalizing sweep of this neoliberal worldview and the underlying logic of control it exhibits. In *The Economic Approach to Human Behavior*, Becker sets out an approach to social analysis built on the premise that "prices and other market instruments" function to "perform most, if not all, of the functions assigned to 'structure' in sociological theories."[45] This approach, Becker argues, is "uniquely powerful because it can integrate a wide range of human behavior." Rather than focusing on material commodities such as oranges or automobiles, Becker writes that the stable preferences that his approach attributes to individuals and groups refer to "underlying objects of choice that are produced by each household using market goods and services, their own time, and other inputs." In Becker's formulation, the human actor receives various economic choices as inputs, processes these inputs against a program of stable preferences, and outputs information that can be used to make "predictions about responses to various changes."[46] This conceptual process of digitization and compression, which extends economic theory beyond direct production and into wider expanses of social existence, can be seen to intensify in Becker's theory of human capital, described by Foucault in *The Birth of Biopolitics* as both extending economic analysis "into

a previously unexplored domain" and "giving a strictly economic interpretation of a whole domain previously thought to be non-economic."[47]
Under this vision of the world, the development of physical and mental
capacities becomes the expression of the individual actor's capacity to gain
the highest possible level of capital in exchange for their activity. Getting
fit, eating well, getting an education, developing social capacities, selecting friends and partners: these things are no longer to be understood as
distinct from value-producing labor. Instead, they become intelligible as an
investment. All activity thus becomes relatable to the chain of production,
and the development of physical, mental, and social capacities becomes a
form of entrepreneurship. As Foucault puts it, this interpretation is less a
description of labor power than it is a "conception of capital-ability which,
according to diverse variables, receives a certain income that is a wage, an
income-wage, so that the worker himself appears as a sort of enterprise for
himself."[48]

The digital-communicational aspects of the economic approach to social
behavior and the theory of human capital are writ large in Becker's work.
In particular, the conceptualization of the human actor as a "decision unit"
that processes inputs against a stable set of preferences points to a diffusion
of the relationship between activity as input and value as output that Marx
describes as production.[49] It is telling that Foucault describes the system of
preferences and decisions configured by neoliberal economic theory as a
form of programming. This form of economics, Foucault writes, "is no longer the analysis of the historical logic of processes"; instead, "it is the analysis of the internal rationality, the strategic programming of individuals'
activity." Society thus becomes intelligible as a kind of computer network,
with an "economy made up of enterprise-units" and a "society made up of
enterprise-units" that appear "at once" as "the principle of decipherment
linked to liberalism and its programming for the rationalization of a society
and an economy."[50]

The principles of control that sit behind this neoliberal worldview—in
which computational concepts (input, "decision unit") allow for the conceptualization of society as a communication system subject to symbolic
capture and regulation—are already visible in a 1945 essay by the economist and philosopher Friedrich Hayek in which one can read that it is
"more than a metaphor" to describe the price system as a "system of tele-
communications" that "enables individual producers to watch merely the
movement of a few pointers, as an engineer might watch the hands of a few
dials, in order to adjust their activities to changes of which they may never
know more than is reflected in the price movement."[51] In its successive

compression of (1) social activity into communication and (2) communication into value, this passage foregrounds the ways in which an expansion of economic thought to the internal workings of individual actors (thus reconfigured as dividuals), and therefore to sociality in general, is built on mediatic metaphors that are believed to be "more that metaphors," supports new processes of organization and value production, and undergirds the new political paradigms that form to support these processes. For Foucault, the major difference between neoliberal analysis and Marxism lies in the former's approach to labor as a concrete rather than abstract phenomenon. But if one builds on the computational-communicational metaphors present in Hayek and Becker as well as in Foucault's own analysis and takes into account the movement from individual to dividual that Deleuze establishes as central to the logic of control, this difference appears as an abstraction of labor at higher levels of granularity, both at the level of the individual-made-dividual and at the level of social interactions. Mapped onto the Marxian system sketched out thus far, the individual or group functions within this model as both laborer and fixed capital, providing both the mechanism for processing inputs into more valuable outputs and the exploitable labor though which this processing takes place.

The question of how cultural practices such as communication might create value and hence constitute a form of labor that can disperse in time and space is central to a number of texts in the Italian *operaismo* tradition, most notably the work of Maurizio Lazzarato and Paolo Virno. Lazzarato, in defining immaterial labor as that which "produces the informational and cultural content of the commodity," points to two distinct movements in the mid-twentieth-century reformulation of work.[52] First, Lazzarato argues, changes to production processes bring about a situation in which the skills involved in direct processes of production have primarily to do with the use of computers: this situation is largely identical to the transformation of work that Agre describes in "Surveillance and Capture" and, Lazzarato suggests, represents a state of affairs in which material practices (work performed on a Ford or Fiat assembly line) are replaced with "immaterial" ones (work performed on a computer). (The reproduction of Lazzarato's distinction between material and immaterial labor here should not be taken as an endorsement of this distinction. The specific power of control is such that its symptoms can be found in critical theorizations as well as in celebratory accounts of specific working practices or modes of organization.) Second, immaterial labor (in Lazzarato's account) takes place when any social actor contributes to "defining and fixing cultural and artistic standards, fashions, tastes, consumer norms ... and public opinion."[53] On the one hand, then, immaterial labor simply accounts for a transformation of the quality of work performed within the physical and temporal boundaries of the workplace. On the other hand, it describes a radical dispersal of value production into all activity that adds value to an object or service, such as the combined attention and discussion required to define a "must-have" item or a "buzz" band. Lazzarato emphasizes this second valence in a later essay titled "From Capital-Labor to Capital-Life," in which he writes of a transformation within the capitalist mode of production after which the purpose of entire sectors of industry is to create worlds rather than (or as well as) to manufacture commodities.[54]

Lazzarato's analysis of the dissolution of boundaries between commodity production and valorization across wider sociocultural fields is extended in Paolo Virno's analysis of post-Fordism. After the communicational mutation of capital, Virno writes, the "ancient tripartitioning" of labor, politics, and intellect is dissolved so that the category of labor comes to absorb "many of the typical characteristics of political action" and "pure intellectual activity."[55] If it is not clear from Lazzarato's conflation of the quite different spheres of factory labor and the general investment of objects with cultural data or from his (not unproblematic) division of material and immaterial practices, Virno's universalizing account underscores the idea that the historical shifts in value production evidenced in the difference between Marx's account of production and those accounts set out in the second half of the twentieth century constitute a transformation of concepts of the human and of social interaction as much as they explain new processes of production and circulation.

The senses, and especially sight, are often theorized as a primary site of the expanded exploitation of human capacities. In his powerful book *The Cinematic Mode of Production*, Jonathan Beller argues that cinema plays a major role in the twentieth-century reformulation of the human as laborer both by reorganizing the relationship between body and commodity and by positing perception and cognition as forms of labor (from the addition of value to an artwork or a celebrity to the function of advertising up to and including the targeted forms perfected by Google). On the first of these theses, Beller writes that "*when we incorporate the image* we ourselves become exchangeable; we have/are social currency."[56] This question of "when" we incorporate the image points to the historical relationship between control and valorization pursued here. One might posit a second historical thread that is intertwined with the one Beller follows: when individuals or groups incorporate the computational metaphor, all processes, whether material or "immaterial," become intelligible as labor.

As Marx observes in the *Grundrisse*, capital can apprehend life only as labor. In the era of industrial production from which Marx writes, the limit of this process is the boundary of the individual; capital apprehends human life *as a whole*, which is to say as an individuated cell of potential energy and, consequently, of value. Theoretical formulations such as human capital and Becker's concept of the human as decision unit, Hayek's telecommunication model of society as price system, Lazzarato's immaterial labor, Virno's dissolution of the boundaries between labor, politics, and intellect, and Beller's attention theory of value address different phenomena and take quite different positions on the political spectrum. Each, though, points to an epistemic shift through which differentiated parts, perceptive states, drives, and desires *within* the individual—all of which can be framed as decisions that indicate some universal productive capacity, consumption-determining preference, or both—become independently subject to capture and valorization, while a far wider range of social phenomena than those limited to direct production processes become constitutive of value. What is critical here is less the accuracy or historical prescience of any one of these individual accounts and more the way in which they collectively point to a distinctive historical production of the human as a network of discrete decisions that appears to obviate the need for any higher level of categorization even as it continues to impose thresholds and conditions of inclusion and expulsion. Such is the logic of control—a logic that, this book argues, is founded on a series of metaphors extending across humans and idealized digital machines.

An Archaeology of Control

To recap the previous parts of the argument advanced here: control is understood as a logical substrate grounding a number of discourses in which the boundary of the workplace as a site of the apprehension and exploitation of the worker's labor disperses, so that life is understood as both uniform and fully representable in digital form, and is thus subject to value extraction across expanded fields of time and space. From the point of view afforded by control, labor is not a compression or digitization of life in certain times and places (the factory, the working day). Instead, all of life—or, at least, all that matters about life—appears already fully digital and thus intelligible as value-creating labor. Concepts such as immaterial labor, cognitive capitalism, and post-Fordist production, as well as the neoliberal vision of the world as a constantly shifting complex of markets, can thus be understood as symptoms of this basic fundamentalism of the digital.

This idealized quantization of life as labor, which represents an expansion of the basic logical formalization affected by capital, is uncannily doubled by the rendering equivalent of all instructions and all data that is an organizing technical principle of any electronic digital computer based on the von Neumann architecture. Far from being a coincidental or anachronistic resemblance, this basic set of equations—the human–social, system–computer complex of metaphors—can be traced through a series of references that indicate the presence of a spectral connection between labor and computational logic in Marx's writings, in particular the four drafts of *Capital* (the *Grundrisse*, the *Manuscript of 1861–63*, the *Manuscripts of 1864–65*, and the three volumes of *Capital* itself). In other words, the progenitors of the fuzzy but operational metaphors between behavior, social organization, and computer technology can be located in both dreams and nightmares of universal subsumption and valorization in political-economic writings of the nineteenth century. Returning to questions concerning the temporality of information-age labor with these connections in mind enables one to look beyond computer technologies as such and to turn to the subtler reorientations of the concept of the worker as a living, thinking being that underpin the economic logic of the present. The nascent possibility of these reorientations, which produce a vision of the material and social world closer to that of the abstract, unbounded logic of capital Mandel finds in Marx than to those evidenced in the limited, concrete practices Marx himself observed in nineteenth-century production, can be located in the work of two of the individuals most commonly associated with the great nineteenth-century precursors of the electronic digital computer: Joseph Marie Jacquard and Charles Babbage. References to both can be found in Marx's writing.

To Jacquard there is only a single, fleeting reference in Marx. It can be found in the *Economic Manuscript of 1861–63*, in a section headed "Division of Labour and Mechanical Workshop. Tool and Machine," where one also finds a discussion of the centrality of the clock to the establishment of the value-form of the commodity and a return to the fundamental difference between tools and machines previously addressed in the *Grundrisse*.[57] In a series of notes on textile production in this section, Marx records a number of passages from the anonymously authored work *The Industry of Nations* (1855), and in one of these passages the innovations of the Jacquard loom and its punch-card system—the automation of detailed weaving, the velocity of work, and the steadiness and lack of susceptibility to jamming or other failures they afford—are detailed in celebratory terms:

"The simple looms are only capable of producing an unfigured fabric, and have no power to form embroidered tissues.... For this purpose a peculiar apparatus is necessary, and looms to which this is attached are called Jacquard looms.... If while the weaving were going forward one or two of the threads of the warp were lifted or depressed while the others were undisturbed, the cloth then made would exhibit a different appearance in that part of it where these disturbed threads were, to the other parts. It would show a certain mark on its surface; and if this disturbance were occasional, these marks would be repeated at a certain distance from one another, and thus a sort of figure would be produced in the cloth. This is what the Jacquard apparatus accomplishes.... Invention of Mr. *Barlow*, exhibited on the Great Exhibition. In this loom, two" (instead of one as previously) "perforated cylinders are used, and the cards are disposed on these in alternate order, so that while one cylinder is in action, the other is changing its card and preparing for work. By this arrangement, the loom can be worked with a velocity 40% greater than that of the ordinary construction. The steadiness of its action also greatly increased, and the strain upon the warp diminished."[58]

When considering the relationship between capital and digitality, it is compelling to think of Marx taking notes from *The Industry of Nations* in the early 1860s, coming upon descriptions of the "methodical" loom in which "certain cards with holes in them," passing "in succession over a square cylinder of wood also perforated with holes, and against which little pieces of wire are pushed at every revolution," lead to the formation of "the most beautiful and delicate designs" the nature of which "entirely depends on the arrangement of the holes in the cards."[59] That Marx's awareness of Jacquard looms becomes apparent across the same pages in which he most plainly states the fundamentally temporal and discrete character of value creation within the productive circuit of capital—as cited earlier, the establishment of a uniform "period in which the value of the commodity" is made equivalent to "the labour time necessary for its production"—is particularly fascinating. Taken alone, this juxtaposition of loom and clock provides nothing more than the (admittedly tantalizing) prospect that Marx might have grasped, even unconsciously, the wider possibilities and implications of digital representation for the conceptualization of labor when honing the theories that would constitute *Capital*. This speculative connection is given greater weight by the wider set of references to Babbage found in Marx's writing from *The Poverty of Philosophy* to the *Grundrisse*, the 1861–1863 manuscripts, and *Capital* itself.

Marx cites Babbage six times in *Capital*, volume 1: two citations appear in the chapter "The Division of Labour and Manufacture," and four appear in the subsequent chapter "Machinery and Large Scale Industry."[60] Each of these references is to Babbage's book *On the Economy of Machinery and Manufactures* (1832), and most of them concern data relating to the improvements in efficiency and value production made possible by specific applications of machine technology to the division of labor. Of greater interest, however, is a single quotation from Babbage that persists throughout a series of Marx's writings, appearing first in 1847 in *The Poverty of Philosophy*, then in the *Economic Manuscripts of 1861–63*, and finally in volume 1 of *Capital*.[61] The passage to which Marx keeps returning concerns the concept of the machine as the ultimate expression of the division of labor: in it Babbage writes "When each process has been reduced to the use of some simple tool, the union of all these tools, actuated by one moving power, constitutes a machine."[62] Here it is possible to identify a clear precursor to the concept of the machine as sociotechnical assemblage that appears in the "fragment on machines" of volume 1 of *Capital*, and that informs Deleuze and Guattari's concept of the machinic assemblage as well as much of the critical writing on post-Fordism.[63] But to limit the significance of Babbage's work on machines to the way in which they congeal practices, ideas, and knowledge in dead labor is to miss the ideal behind Babbage's project on political economy. In the preface to *On the Economy of Machinery*, he writes that "[t]he present volume may be considered as one of the consequences that have resulted from the Calculating-Engine, the construction of which I have been so long superintending. Having been induced, during the last ten years, to visit a considerable number of workshops and factories, both in England and on the continent, for the purposes of endeavouring to make myself acquainted with the various resources of mechanical art, I was insensibly led to apply to them those principles of generalization to which my other pursuits had naturally given rise."[64] In other words, the studies presented in *On the Economy of Machinery* were directly motivated by Babbage's desire to build computing machines, and his design of computing machines was directly motivated by the prospect of better applying the division of labor to human activity.

As the historian of economics Philip Mirowski has noted of Babbage, "The major theme of the *Economy of Machinery* is a revision of Adam Smith's account of the division of labor, where for Babbage the great virtue of dividing labor is separating out the dreary low-skill components of any job and lowering the pay commensurately for those unfortunate enough to qualify to do it."[65] The connection between the division of labor and the principles of computing machines articulated by Babbage thus emerges as a central concern of his political-economic writing and, by extension, a hidden presence in the later critical accounts derived from it by Marx. As Mirowski puts it, "the very architecture of the Analytical Engine, now acknowledged as the first stored-program computer ... constituted a projection of a more perfect factory."[66] Babbage's concept of divided labor is thus nothing less than a protocybernetics under which the modularized body of the individual worker and the divided body of workers, applied to some efficiently divided task, are understood as fundamentally interchangeable with the idea of a digital computer. It is also notable for its intensification of the exploitative and exclusionary aspects of capitalism, whereby computation aids in more efficiently confining "dreary" tasks to those lacking the "skill" for higher pursuits, with wages driven down accordingly. If one wanted to locate a historical precursor to the contemporary tech giant's proclivity for promising always-connected, unbounded, creative lives to enlightened users while at the same time contracting device production to manufacturing companies such as Foxconn, one would need look no further.

In addition to underscoring the fascinating possibility that the critique of capital that rests in large part on the division of labor in Marx is at least in part built on a theory of the latter that conflates factory and computer, the analogy between divided labor and computation in Babbage also provides a precursor to the valorization of cognitive activity that is a precondition for the concept of "information work." The roots of this logic can be found in chapter 19 of *On the Economy of Machinery,* where Babbage writes that "the division of labour can be applied with equal success to mental operations, and ... ensues, by its adoption, the same economy of time" and that "arrangements that ought to regulate the interior economy of a manufactory ... are capable of being usefully employed in paving the road to some of the sublimest investigations of the human mind."[67] Here the computer's logical function provides a model not only for the ideally efficient factory but also for the conscious workings of the human mind—a notion of real subsumption that goes beyond the oft-repeated idea, extracted from the "fragment on machines," that the valorization of thought would emerge through its congealing in inventions and productive technologies.

In *On the Economy of Machinery,* Babbage's account of divided mental labor is confined to the production of mathematical tables (such as logarithm tables), first by mathematicians divided by ability so that any overseer of such work can "avoid the loss arising from the employment of an accomplished mathematician in performing the lowest processes of arithmetic" and then by the development of computing machines that can carry out these processes automatically.[68] These examples remain practically tied to the concept of labor time divided by way of the clock, but in Babbage's interrelated notions that the computing engine could be "usefully employed in paving the road to some of the sublimest investigations of the human mind" and that mental labor could somehow be valorized through analogy with machines, it is possible to identify the roots (and thus the fundamental principles) of control as being densely intertwined with the roots of the labor theory of value.

The equivalence and exchangeability of both actions and thoughts effected by Babbage's mapping of human potentials onto machine actions suggest a historical connection between digitality and capital that precedes the discourse of post-Fordism, immaterial labor, the information economy, and so on. From the division of labor to the discretization of the body by mechanization (which would later be formalized and expanded in Taylorist scientific management) to the conflation of factory and computer and the ideal of thinking as divided labor, one can observe the genealogy of the control society's productive dividual in close proximity to the logic of capital. The specific valorizing logic of control can thus be understood according to the following proposition: if labor under capital is always already digital, then the digitization of practices not formerly understood as labor—communication, sociality, identity, formation, attention—forms a necessary precondition for their conceptualization as such. In other words, *digitization* is a precondition for *subsumption*. If one posits Babbage's worldview as a proxy for capital's optimizing gaze, the conceptual ground for this principle of universal valorization can be located in the ontology expounded in his *Ninth Bridgewater Treatise*. In this text, Babbage states that his "views concerning the laws of nature were greatly enlarged" by his consideration of the calculating engine, before going on to posit the world as fundamentally digital, with God as a grand programmer and emergent phenomena such as evolution and apparent miracles the products of partial equations.[69] Anthony Hyman clarifies this view in his biography of Babbage, writing that

Now the calculating engines—and also, of course, a modern computer—could easily be programmed to proceed according to one law for any number of operations and then proceed according to some other law, the change in operations being programmed *ab initio*. Similarly, reasoned Babbage, the changes in natural law, as evidenced by the creation of new species, were not proof of Heavenly intervention but could also have been programmed by the Creator *ab initio*: that is to say at the time of the Creation. In a similar manner miracles appeared as singularities in the Celestial Program: a miracle was merely a subroutine called down from the Heavenly store. To make the concept more comprehensible to his contemporaries Babbage explained it in terms of singularities in equations of the fourth degree, but he was really thinking in terms of his beloved engines.[70]

This view of the universe as fundamentally digital foreshadows the optimal (and optimizing) worldview of control. Following this, the emergence of control might be understood not as a historical break but rather as an

Repetitions of Process.	Move-ments.	Clock A. Hand set to I.		Clock B. Hand set to III. First difference.	Clock C. Hand set to II. Second difference.
	Pull A.	A. strikes 1	TABLE.	. . . .	. . . .
1	— B.	The hand is advanced (by B.) 3 divisions . .		B. strikes 3	. . . .
	— C.	. . . .		The hand is advanced (by C.) 2 divisions . .	C. strikes 2
	Pull A.	A. strikes 4		. . . .	. . . .
2	— B.	The hand is advanced (by B.) 5 divisions . .		B. strikes 5	. . . .
	— C.	. . . .		The hand is advanced (by C.) 2 divisions . .	C. strikes 2
	Pull A.	A. strikes 9		. . . .	. . . .
3	— B.	The hand is advanced (by B.) 7 divisions . .		B. strikes 7	. . . .
	— C.	. . . .		The hand is advanced (by C.) 2 divisions . .	C. strikes 2
	Pull A.	A. strikes 16		. . . .	. . . .
4	— B.	The hand is advanced (by B.) 9 divisions . .		B. strikes 9	. . . .
	— C.	. . . .		The hand is advanced (by C.) 2 divisions . .	C. strikes 2
	Pull A.	A. strikes 25		. . . .	. . . .
5	— B.	The hand is advanced (by B.) 11 divisions .		B. strikes 11	. . . .
	— C.	. . . .		The hand is advanced (by C.) 2 divisions . .	C. strikes 2
	Pull A.	A. strikes 36		. . . .	. . . .
6	— B.	The hand is advanced (by B.) 13 divisions .		B. strikes 13	. . . .
	— C.	. . . .		The hand is advanced (by C.) 2 divisions . .	C. strikes 2

M

Figure 1.1

On the division of mental labour. *Source*: Charles Babbage, *On the Economy of Machinery and Manufactures* (London: Charles Knight, 1832), 161.

intensification of epistemic conditions to allow an already-existing tendency of capital—the dream of extending valorization to practices beyond time spent in factories or offices—to be realized in concrete social relations. From here the important remaining question concerns the way in which the epistemic conditions for the discretization of the labor process—with its associated breaking down of bodily potentials into certain fixed, predetermined, and optimal series of movements—expands to allow for the conceptual discretization of thought and the body at all times and in all spaces. This expansion must be understood as a historical process that reveals capital's tendency toward a vision of perpetual monetization, a world picture of valorization—in fact, less a world picture than a world model—that would no longer be limited to those hours nominally set aside and bracketed as labor time. Perhaps Babbage also foresaw this diffusion and flexibilization of labor time in his vision of merging universal computation and political economy: in 1827 he sent his two eldest sons, Herschel and Charles, to Bruce Castle school, a utilitarian project in which students gained credit in the form of counters in exchange for "work of any description, done at any time," with the intention that boys would be driven to self-government through the valorization of a wide range of activities.[71] The logic of flexible labor time behind this system was emphasized by the tallying of rewards and punishments on a supporting economic system under which units of time to be redeemed as holiday or used to "pay off" debts incurred as punishments functioned, according to Bernhard Siegert, as a "general equivalent."[72] Rowland Hill and Matthew Hill, the founders of Bruce Castle, describe the system thus: "[Primeal marks] can only be obtained by productions of the very best quality, and, unlike the penal marks, are strictly personal; that is, they cannot be transferred from one boy to another: with a certain number of them, a boy may purchase for himself an additional holiday, which can be obtained by no other means; and in the payment of penalties they may be commuted at an established rate for penal marks."[73] This conceptualization of time as a currency as well as a discrete measure of duration stands as a remarkable precursor to the flexible labor time of late capitalism allegorized in dystopian science fiction from Harlan Ellison's "'Repent, Harlequin!' Said the Ticktockman" (1965) to the film *In Time* (Andrew Niccol, 2011). Whether this temporal flexibility directly fed into Babbage's dream of a digital world of full subsumption is unclear, but in any case it emphasizes the existence of concepts germane to the logic of control in niches and isolated practices before the development of the electronic digital computer—a set of possibilities internal to the logic of capital, awaiting an epistemic grounding.

In summary, where labor-dividing practices from the "classical" factory to Taylorist organization discretize labor processes through a focus on individual bodies and their movements, the techniques of control seek to discretize a much wider range of emergent behaviors, including cognition and social interaction. Techniques for the valorization of such activities rest on the reconceptualization of nominally immaterial phenomena (such as cognition) and nominally material practices (such as bodily activity itself) as digital communication. The forms of violence specific to control thus either consist of or result from the introduction of a break into the realms of both individual human activity and social formations, so that behaviors, affects, and capacities appear as either (1) productive and representable (information) and thus deemed existent or (2) nonproductive and unrepresentable (not information) and thus denied existence. The historical processes through which individuals and social groups become conceivable and representable as digital systems and thus become subject to the violence of selection or expulsion can be apprehended in a nascent form in the use of proto–computing machines for the statistical representation of populations in the late nineteenth century.

Cultural Control

While the preceding analysis makes a case for the understanding of control as a logical mode through which ever-expanding fields of human activity can be conceptualized as labor, it is essential to pause at this point to reiterate the fact that the logic of selection and definition that undergirds digitality is necessarily also a logic of exclusion, a passing on of the malign work of essentialism from the level of appearance to the supposedly more objective level of informatics. Tara McPherson's insistence that "certain modes of racial visibility and knowing coincide or dovetail with specific ways of organizing data" needs to be taken seriously, and might be expanded beyond her focus on the "lenticular logic" of software culture from the 1960s on.[74] Just as the contemporary notion of information economy can be identified as a late-twentieth-century instantiation of an earlier dream (or nightmare) of full valorization, the absolutely inseparable logic of grouping bodies as discrete arrangements of binary symbols—a logic that is clearly founded on older modes of raced, gendered, and classed definition—must also be accounted for within the social, political, and cultural systems of control. As Beller puts it, whenever one considers the many valences of digital production and management, one "might also consider the ways in which the new domain of politics overlays the old, keeping those who were once imagined as 'women,' 'natives,' and/or 'proletarians' in their planet of slums."[75]

The historical imbrications of digitality and the symbolic rendering of bodies come into particular focus with the deployment of electrical machine tabulation in the eleventh US Census of 1890. Herman Hollerith, who planned, designed, and developed the tabulating machines for the 1890 census, would go on to form the Tabulating Machine Company, one of four companies that would merge into the Computing Tabulating Recording Company (CTR) in 1911; a name change in 1924 would rebrand CTR as IBM.[76] Hollerith was evangelical about electrical machine tabulation and published a number of articles setting out his method and its benefits. This method was based on the use of punch cards to represent individuals, with specific sections of the card representing certain characteristics—sex, race, nativity, employment status, marital status, and so on. Perhaps the most concise explanation of Hollerith's system appears in "The Electrical Tabulating Machine," an article published in the *Journal of the Royal Statistical Society* in 1894:

[The] system of electrical tabulation may perhaps most readily be described as the mechanical equivalent of the well-known method of compiling statistics by means of individual cards, upon which the characteristics are indicated by writing. As it would be difficult to construct a machine to read such written cards, I prepare cards by punching holes in them, the relative positions of such holes describing the individual. In the United States Census we used cards of 3¼ inches by 6⅝ inches, the surface of which was divided into 288 imaginary spaces ¼ inch square. To each of these spaces some particular value or meaning is assigned; a hole in one place meaning a white person, in another a black. Here a hole means a certain age-group, there it gives the exact year in that group. A combination of two holes in another part of the card indicates the occupation of the particular individual.[77]

The digitization of the subject for the purposes of bureaucratic management effected by this method is clear enough in Hollerith's account. Writing is replaced by the non-figural mark of the punched hole, and the card (and, by extension, the person rendered on the card) becomes machine-legible. An earlier description of the method, published by Hollerith in 1889, reveals a number of additional valences—most significantly the compression of social reality that the method requires, the inscription in machine language of essentialist conceptualizations of sex and race it performs, and the conflation of machine and social assemblage that it produces:

As the cards are punched they are arranged by enumerators' districts, which form our unit of area. The first compilation that would be desired would be to obtain the statistics for each enumeration district according to some few condensed groupings of facts. Thus it might be desired to know the number of males and of females, of native born and of foreign born, of whites and of colored, of single, married, and widowed, the number at each of centre groups of ages, etc., in each enumeration district. In order to obtain such statistics the corresponding binding-posts on the back of the press frame are connected, by means of suitable piece of covered wire, with the binding-posts of the counters upon which it is desired to register the corresponding facts. A proper battery being arranged in circuit, it is apparent that if a card is placed on the hard rubber bed plate, and the box of the press brought down upon the card, the pins corresponding with the punched spaces will close the circuit through the magnets of the corresponding counters which thus register one each. If the counters are first set at 0, and the cards of the given enumeration district then passed through the press one by one, the number of males and of females, of whites and of colored, etc., will be indicated on the corresponding counters.[78]

In this extract, from an article written several decades before the emergence of communicative or immaterial labor, the principle of automating a digital model of the social is seen to be inseparable from essentialist configurations of identity. That the Hollerith method is based on conflating the material properties of a machine and the defining characteristics of a human—so that holes in punch cards stand for the predicates that come to define a given individual, and a power source, wires, and counters are put to work to express these predicates in preferred combinations—underscores the way in which the abstraction of concrete technologies into models for apprehending social reality provides the grounding logic of control as it is defined in this book. The centrality of race and nativity to the examples and diagrams Hollerith provides in the 1889 article also makes it clear that digital social logic in its earliest forms was already premised on a reformulation of old essentialisms under the new banner of apparently objective data collection and management.

The example of Hollerith's electrical tabulation system is not presented here to support an argument that the computer is inherently racist or sexist, nor is the earlier discussion of Babbage intended to suggest that computers are inherently capitalist. Rather, these examples from the prehistory of the digital computer are intended to foreground the ways in which deployments of the computer as a tool for governance can produce metaphors that normalize and legitimate real practices of essentialist definition and exclusion.

Beyond the ease with which Hollerith's 1889 article maps identity onto digital representation and management, it is also worth noting a more complex association it makes between this representation and its wider aesthetic situation that is fascinating in its prefiguration of the cultural logic of control. In making his case for the importance of gathering statistical

FIG. 8.

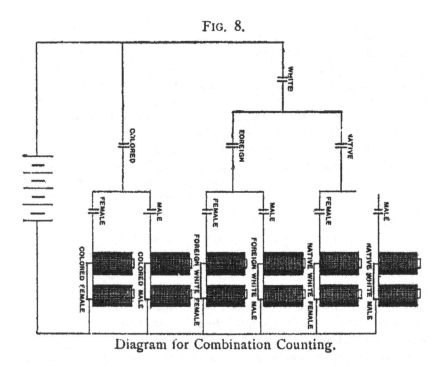

Diagram for Combination Counting.

Figure 1.2
Diagram for combination counting of race and nativity. *Source*: Herman Hollerith,
"An Electrical Tabulating System," *Columbia University School of Mines Quarterly* 10.16
(1889): 254.

data about a society, Hollerith utilizes an extended analogy between data
collection, on the one hand, and interpretation, the objective conditions
of society, and photography, on the other, noting that "the enumeration
of a census corresponds with the exposure of the plate in photography,
while the compilation of a census corresponds with the development of the
photographic plate."[79] In Hollerith's extraordinary analogy, thorough planning and effective data collection for the purpose of a census are directly
equivalent to a technically correct photographic process: listing the categories included in the previous census of 1880, Hollerith states that "such
an enumeration as this, if made thoroughly ... would scarcely be termed
under-exposed."[80]

Race or color: whether white, black, mulatto, Chinese, or Indian.
Sex.
Age.
Relationship of each person enumerated to the head of the family.

Civil or conjugal condition: whether single, married, widowed, or divorced.
Whether married during the census year.
Occupation.
Number of months unemployed.
Whether sick or otherwise temporarily disabled so as to be unable to attend to ordinary business or duties on the day of the enumeration; what was the sickness or disability?
Whether blind, deaf and dumb, idiotic, insane, maimed, crippled, bedridden, or otherwise disabled.
Whether the person attended school during the census year.
Cannot read.
Cannot write.
Place of birth.
Place of birth of father.
Place of birth of mother.[81]

Where, for Hollerith, data collection across these categories is equivalent to the act of photographic capture, the tabulation of these data to show certain correlations and trends is equivalent to photographic development—an analogy that emphasizes the use of the term *processing* across both photography and data management:

As the first flow of the developer brings out the prominent points of our photographic picture, so in the case of a census the first tabulations will show the main features of our population. As the development is continued, a multitude or detail appears in every part, while at the same time the prominent features are strengthened, and sharpened in definition, giving finally a picture full of life and vigor. Such would be the result of a properly compiled and digested census from a thorough enumeration. If this country is to expend $3,000,000 on the exposure of the plate, ought not the picture be properly developed?[82]

A photograph, for Hollerith, is objectively the same as the material conditions it captures, and a data model composed of specific, selected categories is equivalent to a correctly developed photograph. Here it is possible to observe the folding of the real and the imaginary into the symbolic, a process that is central to biopolitical management and that would go on to become central to so-called immaterial production. It is also possible to observe the way in which nondiscrete and noninteractive representational forms are shown to remain central to the representational and productive modes of the digital worldview, if only as orienting and normalizing interfaces.

Remarkably, a comparable analogy between digital representation and photography is given some fifty-nine years later in Norbert Wiener's foundational book on cybernetics (first published in 1948), albeit to suggest that informatic modeling *cannot* be applied to actually existing societies in any meaningful way. Wiener writes that "[f]or a good statistic of society, we need long runs *under essentially constant conditions*, just as for a good resolution of light we need a lens with a large aperture," suggesting that the fluctuating character of material history renders it a "very poor testing ground" for statistical analysis.[83] Regardless of these objections, the conflation of discrete data and analog photography found in both Hollerith and Wiener is isomorphic with the supposed immateriality of software that undergirds the production of vapor theory in commercial and critical analyses of computation as well as in socioeconomic theories and practices of control-era capitalism. It is only by suppressing the materiality and necessary limitations of given technologies—as when conflating an ideal camera (with a lens *"made of a material so homogeneous that the delay of light in different parts of the lens conforms to the proper designed amount by less than a small part of the wavelength"*[84]) with an ideal computer, which Wiener rejects as a workable mode of social representation—that these technologies can serve as models for the management of the material world. One cannot but think here of Vilém Flusser's formulation of the camera as a "prototype of the apparatuses that have become so decisive for the present and the immediate future," ranging from apparatuses that "assume giant size, threatening to disappear from our field of vision (like the apparatus of management)," to those that are "microscopic in size so as to totally escape our grasp (like the chips in electronic apparatuses)."[85] As Hollerith's account of his tabulation system shows, the conflation of apparently distinct apparatuses is central to the material and cultural-political processes of management, exploitation, and exclusion that ground digital representations of the social.

The cultural and technical logics that informed mechanical census tabulation in the late nineteenth century remind one that cultural politics are always implicated in socioeconomic and cultural impositions of control. With this in mind, it is worth reiterating the reasons why it remains necessary to delineate an aesthetics as well as a politics of digitality: on the one hand, the conditions of conceptualization and representability that undergird control-era capital constitute a totalizing aesthetic mode that is inseparable from the systems of representation that sustain and intensify economic and cultural inequality through a strict logic of definition and selection; on the other hand, the formation and maintenance of this representational mode is premised on the volitional activity of social actors, and thus requires the production and distribution of images and narratives. It is the relationship between these representational modes that lies behind the current formation of control as an episteme of intertwined technical, social, and cultural logics.

In order to provide a materialist historical basis to this theoretical peri-odization of control, one must now turn to a post–World War II moment in which a number of its grounding ideas and practices assumed central roles in both technological and socioeconomic thought.[86] It is in this period that the word *cybernetics* came to signify both an intensification of research into self-regulating machines and the emergence of programs of thought orga-nized around computational metaphors.

Although the constellation of knowledge and practices named "cyber-netics" (it is difficult to define it as a discipline) has received critical atten-tion as the end of philosophy, a radical epistemology, an epoch in the history of the social sciences, and a utopian project bound up with coun-tercultural movements,[87] it has been less commonly analyzed as a moment in the history of political economy or as the epistemic grounding for a worldview that posits all material objects and their interactions as digital and thus predisposed to exchange and valorization. In its location of cyber-netic logic across scientific discourse and literary texts, N. Katherine Hay-les's book *How We Became Posthuman* represents an indispensible precursor to such an inquiry. However, where Hayles is principally concerned with the epistemological implications of cybernetics (its separation of informa-tion from material body) and with literary representations of the kinds of futuristic technological augmentation that might be imagined to emerge out of it (such as Bernard Wolfe's novel *Limbo* and Philip K. Dick's writing), the present book is concerned with the subtler, less spectacular ways in which cybernetic principles came to expand from specialist knowledge into forms of *doxa*. In short, the following chapters emphasize the epistemic power of digital logic rather than the signature technologies and projected futures with which it is often associated.

In the 1940s and 1950s, research, debate, and practice surrounding com-puting machines and cybernetics drew together a range of earlier develop-ments in computation (such as Babbage's work and the use of Hollerith's tabulating machines in the 1890 US census, as well as the analog computers of Vannevar Bush and others) and social thought (not only the Babbage work analyzed earlier but also developments such as Frederick Lanchester's inauguration of operations research with his differential "power laws" of 1916[88]). The interdisciplinary movements of this period, set against the "uninterrupted" and mutating narrative of capitalism, foreground ques-tions of how and why complex, time-critical processes of social interac-tion might be not only processed but also understood, represented, and projected into the future as symbolic systems—that is to say, digitally. In the diffusion of this logic into the dominant socioeconomic paradigms of the present, one can observe the cunning of history at work—but not in the subsumption of progressive political ideals within capitalist accumulation that Nancy Fraser identifies within certain feminist movements after the 1960s.[89] The notion of the cunning of history is powerful not only because it emphasizes the ways in which capital subsumes radical principles, but also because it evokes the manner in which capital expands its purview—not through intentional, long-term plans but through multiple, overlapping, ad hoc responses and adoptions. The particular instance of the cunning of history pursued here is manifested not in the co-option of ideals of equality but rather in the apparently seamless application of specific technical principles to the general fields of human thought, activity, and interaction.

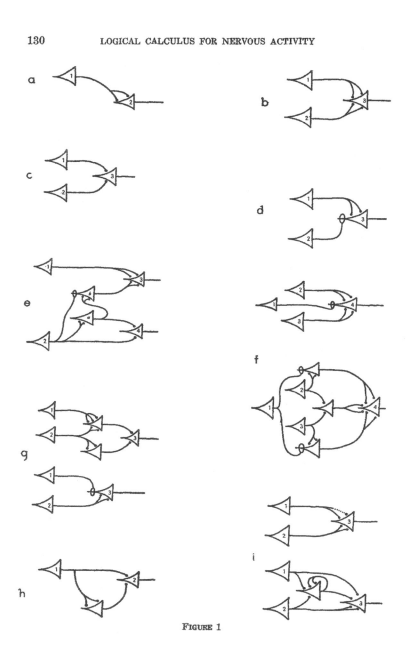

FIGURE 1

Figure 1.3

The McCulloch–Pitts neuron. *Source*: Warren McCulloch and Walter Pitts, "A Logical Calculus of the Ideas Imminent in Nervous Activity," *Bulletin of Mathematical Biophysics* 5 (1943): 130.

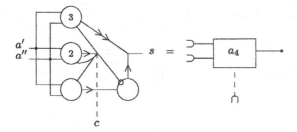

Figure 1.4
Influence of the McCulloch–Pitts model of neural activity on the design of electronic
digital computer architecture. *Source*: John von Neumann, "First Draft of a Report on
the EDVAC," Contract No. W-670-ORD-4926 between the US Ordnance Department
and the University of Pennsylvania, 1945, 10.

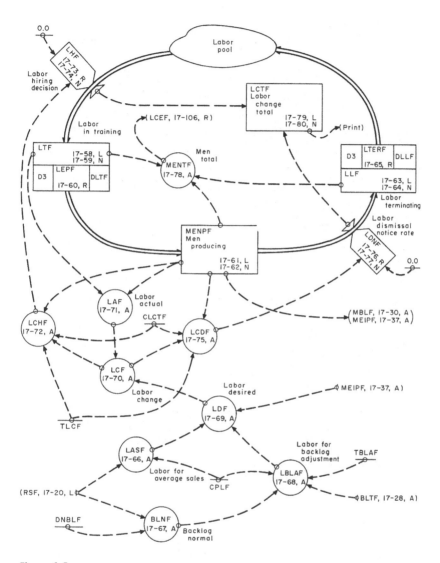

Figure 1.5

Labor flow. *Source*: Jay W. Forrester, *Industrial Dynamics* (Cambridge, MA: MIT Press, 1961), 230. Permission granted by the author.

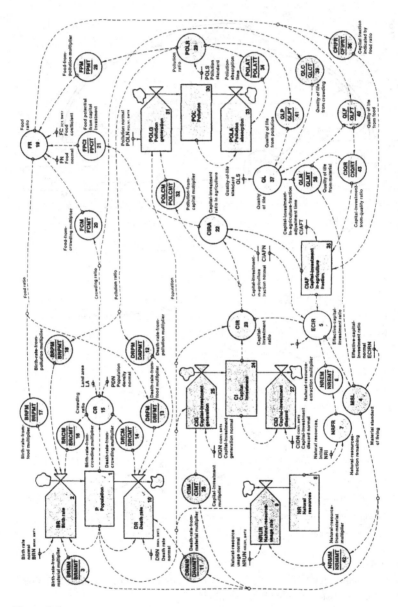

Figure 1.6
System-dynamic model of world population change and economic growth, 1971–
2021. Source: Jay Forrester, *World Dynamics* (Cambridge, MA: Wright-Allen Press,
1971), 21.

2 Cybernetics, or The Digitization of the Social

The forms of capitalist production accede to representation in each epoch ... by mobilizing concepts and tools that were initially developed largely autonomously in the theoretical sphere or the domain of basic scientific research. This is the case with neurology and computer science today.

—Luc Boltanski and Ève Chiapello, *The New Spirit of Capitalism*

History and Historicism

In "War against the Center," Peter Galison calls for a move beyond culturally led periodization theories such as Jameson's postmodernism thesis and toward studies of the concrete historical situations that bring about what are often understood as broad cultural changes. From its opening methodological preamble onwards, Galison's critique gestures toward the conflation of digital technology and cultural-historical totality that is defined here as the cultural logic of control. He notes that although "even the briefest of web searches yields hundreds of sites with titles like 'Internet=Postmodernism,'" an answer to the question of *how* we lurched from a "centered modernism" to an "aesthetic, architectural, economic, and, according to some, metaphysical placelessness" is "less clear." Galison argues that the "intellectual, pragmatic, aesthetic, and stochastic drives" implicated by cultural-Marxian notions of base–superstructure reflection, while not necessarily useless for the analysis of cultural-historical transformations, are overly "abstract." In order to move away from these abstractions, Galison proposes to address the emergence of decentralization not through economic shifts ("1973 with the oil crisis and subsequent economic upheaval") or broad cultural turns ("the social upheavals or deconstructivist literary-theoretical work of the 1960s") or specific technological developments (the Internet), but by starting from "bombs"—specifically, the activities of the U.S Strategic Bombing Survey founded in 1944.[1] The confluence of military, logistical, urban-planning, economic, technological, and critical-theoretical ways of seeing in Galison's approach to postmodernity makes it clear that he is concerned less with evoking the Strategic Bombing Survey as a starting point for the processes of distribution so often asserted as definitive of current geopolitical and economic organization than with evoking the concrete event as a specific instance within a complex of intersecting historical currents. This chapter takes Galison's essay as inspiration for an inquiry into the ways in which conceptual frames developed in relation to specific scientific, technological, and military practices can slip into nominally unrelated fields of social and political organization.

As Galison demonstrates, the Strategic Bombing Survey was an "immense," multidisciplinary operation that deployed operations research methods to predict the optimal targets through which allied forces could bring about the "collapse of the German economy as a whole."[2] The range and background of the participants make clear the breadth of disciplinary insights the survey was designed to mobilize:

The head of a major mining firm directed work on munitions, the executive vice president and general manager of Standard Oil directed the petroleum division, and a former vice president of the Curtiss-Wright Corporation ran the Aircraft Division. Appropriately enough, Franklin d'Olier, president of Prudential Insurance, ran the whole of the Survey—the greatest damage-assessment program in history. Among the major figures running other divisions were John Kenneth Galbraith (overall economic effects), George Ball (transportation), and Paul Nitze (equipment and utilities). Starting on the lower rungs of the ladder were Marxist economist Paul Baran and poet W. H. Auden.[3]

Statistical analysis in the aftermath of the bombing raids planned through this multifaceted program of operations research, Galison observes, produced a simple but telling insight into bombing strategy and, by extension, urban planning: aerial warfare "worked when it hit concentrated, centralized production standing at a functional node, upstream of many other industries," and "failed when the target nation effectively dispersed their factories [as with the production of tertraethyl lead, essential for the production of aviation fuel]."[4]

The major insight of the Strategic Bombing Survey—that a decentralized structure was in many situations (especially that of total war) stronger than a centralized one—provides, for Galison, a crucial moment in the emergence of the decentralized and self-reflexive worldview later characterized as that of postmodernism. After the war and especially after the allies had analyzed the rubble of Hiroshima, Nagasaki, and Hamburg, the survey's insights were brought to bear not only on the visualization of future enemies but also for the urban planning, transport systems, and, later, communication infrastructures of the United States itself. Galison suggests that although "the bombsight eye had already begun to reflect back" before Hiroshima, it was in the postwar applications of the Bombing Survey's operations-research logic to "each community, each industry, each factory" that the "remarkable practice of training Americans to see themselves as targets" led to widespread programs of decentralization in urban planning, industrial management, architecture, and communications design, among any other fields.[5]

As compelling and methodologically instructive as Galison's argument is, there is no great value in simply rehearsing all of its nuances and historical insights here. There is, however, significant value in exploring a logical conundrum that crops up toward the end of "War against the Center" and that points to the methodological concerns required to address the concept of control as episteme examined throughout this book. In proposing some of the broader implications of the Strategic Bombing Survey in the years following World War II, Galison turns to the interdisciplinary field of cybernetics: "One thinks here of the origins of cybernetics, launched when Norbert Wiener began to think of the enemy bomber pilot as a kind of feedback machine that could be mimicked electronically; from there, it was a short step to thinking of the Allied gunner in the same way. Then human physiology began to appear as a cybernetic system, then the human mind, then life, *then even the world system as a whole.*"[6] There is nothing controversial in Galison's location of the roots of cybernetics in World War II ballistics research or in the range of fields that the same research came to bear on in the decades after the 1940s. Both of these historical threads are well documented in primary literature and have received substantial scholarly attention.[7] Within the context of Galison's argument, however, a conceptual slippage essential to the formation of the control episteme can be discerned. Galison begins "War against the Center" by proposing to explain a cultural-theoretical abstraction (the cultural logic of late capitalism) through a specific historical event (the Strategic Bombing Survey and the subsequent promotion of its insights within a range of areas in domestic policy); in making the claim about cybernetics quoted earlier, however, he effectively describes a situation in which the logical basis of a concrete historical practice (cybernetics research) blurs into the logical basis of a contemporary sociocultural abstraction (the "world system"—a concept that is as invested in the representational logic of base–superstructure reflection as the notion of the cultural logic of late capitalism is). In turning Galison's materializing trajectory into a loop, this claim does not undercut the logic of his argument so much as it foregrounds the central conundrum behind the emergence of control: How can specific technical practices transmute into a general social and cultural logic?

The cunning of the historical processes implicated in this question is illustrated in the list of phenomena to which cybernetic logic became attached in Galison's account: first, the enemy bomber pilot is modeled as if he is a machine, then the allied pilot is, then animal physiology, then the brain, then "life itself," then the "world system" that can only be understood as a network of capital flows. There is a historical sleight of hand in this apparently straightforward list of applications. Although the first five examples can, however roughly, be reduced to a finite number of variables that can be understood as automatic (that is to say, goal-seeking but preconscious) phenomena constrained by physical and/or technical limitations (the maneuverability of a bomber and the scope of its control mechanism; the structure, density, and conductivity of nerves; the growth and function of neurons; systems of circulation, respiration, excretion, and so on; environmental factors necessitating adaptation), the last category, the world system, is not only subject to such technical determination but is also constituted by the abstract logic of capital, implemented through unconnected human actors' conscious *and* unconscious decisions, invested in certain ideological formations, driven by commitments to private property and free markets, and so on. These phenomena cannot be so simply reduced to the abstracted, synchronic logic of neurons firing or hormones being released that are conceptualized by cybernetics as happening automatically under certain stimuli (or inputs) unless one commits to the principle that the market relations of global capitalism (and, by extension, the forms of violent expulsion, expropriation, exploitation, and subjectification that these relations entail) are fully natural. Despite this contradiction, over the middle decades of the twentieth century, the logic of cybernetics was increasingly applied to fields such as economics and management until the name "cybernetics" itself disappeared and the methods it describes came to constitute a seemingly objective component of political economy and management theory, among many other fields (humanities research being only one further example). If one follows Galison's example, then, it is the question of *how* this happened that is of concern in the historical contextualization of control. Before this, however, a few brief notes on terminology and method are necessary to clear the way for the political-economic inquiry that is at the heart of this venture into the digital cultural logic of the control.

Cybernetics has been described in many ways, but at root it can be thought of as a logical framework for understanding self-regulation in biological life and machines and, by extension, as a logical basis through which biological organisms and machines can be considered formally interchangeable. Many thorough accounts of the definition, history, prehistory, and future of cybernetics exist, and it is not this book's purpose to provide another.

Although the formal (if hard to define) discipline of cybernetics plays a central role in the emergence of new approaches to neural and nervous function, biological adaption and evolution, and psychology, among other disciplines dealing with "automatic" functions that respond to environmental changes, its social applications must be understood as interconnected with a range of other disciplines both through direct causal relationships and shared methodological or conceptual principles. Galison's collective term *Manichean sciences*, which he uses in "The Ontology of the Enemy" to bracket the World War II disciplines of operations research, game theory, and cybernetics, might thus be expanded to account for practices such as systems theory and system dynamics, which, although not in every instance rooted in the same wartime imperatives of determining unknowable enemy behavior, are clearly invested with the same logics of command and control as the heterogeneous complex of practices definable as cybernetics.[8] As chapter 1 argued, a nascent form of this cybernetic logic can be observed in Babbage's political-economic fantasies if not in the logic of capital itself, and a more fully developed and universalized form can be observed behind Hayek and Becker's work and located in the historical formations critiqued by Deleuze, Lazzarato, Virno, and Beller, among others. This expanded notion of cybernetic logic (as a wider set of approaches and practices than those generally defined as falling under the term *cybernetics*) is central to Tiqqun's description of a "cybernetic hypothesis" that expands to determine globalized political economy from the mid–twentieth century on as well as to the historical project attempted in this book.[9] Thus, the term *cybernetic logic* is here deployed to account for a range of practices and methodologies that render the world legible through processes of capture, digitization, modeling, and prediction.

Statistical Forecasting

A glance at the text that is generally considered the milestone in the emergence of cybernetics as an interdisciplinary field immediately reveals some of the complex historical processes at work in the gradual application of cybernetic logic to social phenomena. Wiener's account of the foundations of cybernetics in his 1948 book posits the history of binary logic (from Gottfried Wilhelm Leibniz to Bertrand Russell, David Hilbert, and Alan Turing) and self-regulating mechanisms (from James Clerk Maxwell's 1868 paper "On Governors" on) as progenitors of cybernetics.[10] In the introduction to *Cybernetics*, Wiener posits his work on computing machines for the solution of differential equations (with Vannevar Bush) and his study of electronic networks (with his former student Yuk Wing Lee) as foundational to the development of the cybernetic approach and points to wartime work on self-guiding antiaircraft ballistics systems (with Julian Bigelow) as the moment at which cybernetics as a general study of humans and machines as control–communication systems began to coalesce.[11]

The vector along which the slippage of materialist cybernetics into socioeconomic applications appears to move is that of prediction or, more accurately, forecasting.[12] It is the prospect of modeling and forecasting (and thus valorizing) social behavior that drives the desire for a universally applicable cybernetics, and this prospect must be seen as grounded in the idealization of the universal digital computer touched on in the introduction and chapter 1. This principle of prediction is fundamental to Oliver, Pierce, and Shannon's 1948 work on PCM, in which the advantages of the method (and of binary coding in general) lie in the limits it places on possibility and the increase in representational efficiency and predictability that this limitation entails. Writing on the usefulness of binary coding for technical communication in the presence of noise, Oliver, Pierce, and Shannon assert that this method is optimal in that it most successfully minimizes the chances of errors: "If the received signal lies between a and b, and is closer (say) to b," they write, "we guess that b was sent. If the noise is small enough, we shall always be right."[13] If this claim exemplifies the specific technical uses of cybernetics logic for communication, the connection between technical and universal forecasting centered on the computing machine is both clearer and more universal in Turing's 1950 essay "Computing Machinery and Intelligence," where he states:

The system of the "universe as a whole" is such that quite small errors in the initial conditions can have an overwhelming effect at a later time. The displacement of a single electron by a billionth of a centimeter at one moment might make the difference between a man being killed by an avalanche a year later, or escaping. It is an essential property of the mechanical systems which we have called "discrete state machines" that this phenomenon does not occur. Even when we consider the actual physical machines instead of the idealized machines, reasonably accurate knowledge of the state at one moment yields reasonably accurate knowledge any number of steps later.[14]

Crucially, for Oliver, Pierce, Shannon, and Turing it is this predictability-through-limitation that ultimately *distinguishes* the discretely coded communication system or the computing machine from the "universe as a whole." In the years following the earliest work in cybernetics, however, a series of conceptual shifts toward the identification of universe and computer take place, driven by a desire to apply the predictability of the latter to the representation and management of the former. It is the practices and conceptual structures behind this shift, all of which rest on basic principles developed in relation to biological and artificial hardware while discarding the clear caveat about social applications attached to these principles, that connect the nineteenth-century dreams of political-economic and governmental digitality to the epistemic formations of the control era.

Put simply, cybernetics in its foundational years was clearly directed toward technical machines and human–machine interaction, with little concern for the conscious types of human–human interaction that might lead to the development of systems of socioeconomic modeling, prediction, and control. Although identifying methods of probabilistic modeling applicable to certain functions of organic life is clearly central to Wiener's work with Bigelow and Arturo Rosenblueth, the type and range of this modeling are clearly defined in relation to specific, highly limited conditions. This is emphasized in "Behavior, Purpose, and Teleology," published in January 1943 as a general overview of the concerns that informed the antiaircraft servomechanism work. In this essay, Rosenblueth, Wiener, and Bigelow establish the principles of purposeful behavior and feedback, noting that the latter can be divided into four types: positive feedback (where some element of a system's output rejoins its input channel, as with microphone-speaker feedback); negative feedback (where some element of a systems input causes a change in its output, such as when one adjusts one's stance and hand position based on the trajectory and speed of an approaching ball); nonextrapolative negative feedback (where no prediction is involved—the example given is an "amoeba[,] [which] merely follows the source to which it reacts" with "no evidence that it extrapolates the path of a moving source"); and extrapolative negative feedback (where prediction occurs—the example given is a cat that, in attempting to catch a mouse, "does not run directly toward the region where the mouse is at any given time, but moves toward an extrapolated future position").[15] In each of these examples, feedback is clearly used to describe automatic or reflexive behavior over very short periods of time, rather than to account for the longer periods of social negotiation and contestation that can be defined as history. Even when the cybernetic mode of analysis was initially applied to humans in Wiener and Bigelow's work on the antiaircraft predictor, it was in ways that were clearly incompatible with the kinds of long-term forecasting that would permit socioeconomic modeling.

Although work on computers and networks introduced many of the ideas and diagrams that would be central to later cybernetic principles, Wiener notes that the ballistics project introduced two elements that would distinguish cybernetics from purely technical research: the addition of human agents (the enemy pilot and the antiaircraft gunner) and the imperative of prediction; due to the "speed of the airplane" having rendered "obsolete all classical methods of the direction of fire" by the 1940s, the major concern of this research was the necessity of "predicting the future position of the plane" that was under enemy pilot control so that "missile and target may come together in space at some time in the future."[16] For Wiener and Bigelow, the solution was to do away with the notion of predicting the pilot's conscious mental activity in favor of statistically modeling the most likely outcome of his instinctive behavior under the stress of combat and mediated by the limitations of the airplane's technical systems. The passage of *Cybernetics* in which Wiener discusses the logic behind this principle itemizes these material constraints and is worth reproducing at length:

> If the action were completely at the disposal of the pilot, and the pilot were to make the sort of intelligent use of his chances that we anticipate in a good poker player, for example, he has so much opportunity to modify his expected position before the arrival of a shell that we would not reckon the chances of hitting him to be very good, except perhaps in the case of a very wasteful barrage fire. On the other hand the pilot does *not* have completely free chance to maneuver at his will. For one thing, he is in a pane going at an exceedingly high speed, and any too sudden deviation from his course will produce an acceleration that will render him unconscious, and may disintegrate the plane. Then too he can only control the plane by moving his control surfaces, and the new regimen of flow that established takes some time to develop.... Moreover, an aviator under the strain of combat conditions is scarcely in the mood to engage in any very complicated and untrammeled voluntary behavior, and is quite likely to follow out the patter of activity in which he has been trained.[17]

What this foundational experiment makes clear is that although the early, materialist form of cybernetics (exemplified by Wiener and Bigelow's work) was ostensibly applied to the prediction of human behavior, such prediction was established as practicable only under strict conditions in which the physical constraints imposed by the specific situation, the technical possibilities inherent to a given machine (the airplane), and the mental rigors of a high-stress situation severely limit the field of possible behaviors.[18]

The statistical grounding of prediction (or, more accurately, forecasting) is emphasized by work in the field of communications engineering that represents a central thread of what would come to be called cybernetics. Following his discussion of the antiaircraft predictor in *Cybernetics*, Wiener recounts his work with Bigelow on the "inseparable" fields of control engineering and communications engineering, stating that these fields were predicated on a concept of the message not as a specific, unique, and meaningful entity but as a "discrete or continuous sequence of measurable events distributed in time."[19] Addressing the problem of the mechanical or electrical processing of messages required to transmit and decode them, Wiener notes that he and Bigelow found that the better calibrated a given apparatus was to predict a smooth wave, the more easily it would be thrown into oscillation by "small departures," so the prediction of smooth and rough curves required different apparatuses.[20] The solution Bigelow and Wiener determined was digital and binary: by first establishing a mean square error for a given series, they were then able to "translate the problem of optimum prediction to the determination of a specific operator which should reduce into a minimum a specific positive quantity."[21] From this perspective, Wiener writes, "the transmission of information" would be "impossible save as a transmission of alternatives." As such, the actually occurring event, be it a human action or a transmission of symbols between machines, is nothing more than a selection from a field of interchangeable possibilities—a flattened spatiotemporal logic that is exemplified by the Markov chain, a system for modeling multiple possibilities of equal probability (the most simple being the coin toss) that renders the complex systems produced by multiple selections subject to statistical forecast.[22] This is the same principle that undergirds much of Shannon's mathematical communication theory.

The cybernetic deployment of digitality as a logic that extends beyond the computing machine is perhaps most clearly apparent in Warren McCulloch and Walter Pitts's work on neural activity. Wiener notes that the concepts espoused in "Behavior, Purpose, and Teleology" were disseminated by Rosenblueth at a meeting in New York in May 1942, where an interdisciplinary group consisting of psychiatrists, physiologists, psychobiologists, and anthropologists met under the directorship of Frank Fremont-Smith of the Josiah Macy Jr. Foundation. The purpose of this meeting was to discuss "cerebral inhibition," or the apprehension of mental processes as both automatic and limited by material constraints such as the structure and firing capacities of neurons.[23] Present at this meeting was McCulloch, whose 1943 paper "A Logical Calculus of the Ideas Immanent in Nervous Activity" is now widely credited as a foundational work on neural networks.[24] In "A Logical Calculus," McCulloch, in collaboration with the logician Walter Pitts, builds on Turing's 1936 paper "On Computable Numbers with an Application to the *Entscheidungsproblem*" (as well as on the logical tradition of Russell, Alfred North Whitehead, and Hilbert, among others) to present a simplified model of neurons as "all-or-none" binary switches.[25] Like the Turing machine, which directly inspired them, the McCulloch–Pitts neurons are explicitly idealized, which is to say that they are a metaphor based on a metaphor. Although McCulloch and Pitts's paper is significant for proposing that the same principle of statistical forecasting found in Wiener and Bigelow's anti-aircraft predictor can be applied to the firing of neurons in a network, the authors quite clearly state (as do Wiener and Bigelow) that this application is limited and technical, and that it can deduce "neither afferent from central, nor central from efferent, nor past from present activities."[26]

Despite the explicit insistence on material specificity that characterized this initial research into feedback and self-regulation in aircraft flight, technical communication, and neural activity, the ideal of forecasting human behavior would prove to be seductive enough that the extensive formal limitations and caveats imposed by Bigelow, McCulloch and Pitts, Rosenblueth, Shannon, and Wiener were quickly reasoned away by others. The Macy conferences, held in New York between 1946 and 1953, provide a critical insight into the processes through which the initial, tightly bounded principles of cybernetics were progressively extended in ways that proved remarkably germane to the expansionist logic of postwar capital.

Cybernetics Diffusing

McCulloch's memo to the participants of the fourth Macy conference, prepared and distributed in October 1947 to make up for the lack of documentation produced across the first three conferences, makes clear the initial barriers between the formulation of cybernetics set out by its pioneers (Wiener, Bigelow, Rosenblueth, McCulloch) and the possibility of applying cybernetic methods and vocabulary within the human sciences. McCulloch recalls among the discussions at the first conference material on computing machines and game theory (by von Neumann), synchronization in the strychninized nervous system (by the neurophysiologist Frédéric Bremer), "transacted frogs' olfactory lobes under caffeine" (by the neurophysiologist and behavioral scientist Ralph W. Gerard), goal-seeking automata (by Wiener), and the "reiterative performance of the anatomically damaged brain when presented with such various stimuli as Rorschach cards" (by the clinical psychologist Molly Harrower).[27] Discussions following the presentations by the anthropologists Gregory Bateson and Margaret Mead and the philosopher F. S. C. Northrop, each of whom had been present at the May 1942 meeting, "smoldered on several scores."[28] As McCulloch describes the first conference, the papers on possible social scientific and philosophical applications of cybernetics aroused the following disputes from the outset:

First, whether man's ideas derived from science had aught to do with his judgments of value of his conduct of life other than in technological matters. Second, whether even with man's invention of computing machines and mechanisms embodying purposive behavior man was even now in a position to consider his own doings as a proper object of science in Northrop's sense. And finally—because Northrop had used a distinction I had earlier half-made, between signals reverberating in nervous nets or computing machines, and signs which might be made on paper or alterations of structure within the net, or else either punched records or locked in relays of machines, the latter being only surrogates for signals—whether we were not unduly extending our theory of feedback, regenerative or inverse, into domains of which we were ignorant as to what the significant variables were.[29]

In recalling these tensions, McCulloch's account makes it clear that the participants concerned with hardware (whether biological or artificial) were in the majority at the first Macy conference, and that even those not directly concerned with such matters (such as Northrop) put forth immediate doubts about the applicability of theories developed in relation to biology and computer hardware to the analysis of social groups.[30] McCulloch's statement also evidences the necessary relationship between socioeconomic digitality and essentialism: it is only when one is convinced that one is in full command of all of the significant variables governing a given domain, his account states, that one might attempt to extend theories based on feedback into moral and social inquiry.

The work presented by Bateson and Mead at the first Macy conference is broadly indicative of the way in which the former envisaged cybernetics providing a theoretical basis for the human sciences. McCulloch writes that Bateson and Mead's presentation "centered around a description of communities which achieved stability by inverse feedback"—an area of inquiry that would inform Bateson's 1949 essay "Bali: The Value System of a Steady State."[31] In this essay, as in much of Bateson's later work, the concepts of information, communication, feedback, and homeostasis function as a vocabulary with which to describe a wide range of structures, from particular social formations to the role of art in culture to psychological phenomena and patterns of alcoholism. It should be noted that cybernetics, in this deployment, is nothing more than a *diagnostic* method within the human sciences. It provides a vocabulary and a set of perspectives for describing already-existing behaviors and social structures. The question of applying the methods that produced this diagnostic vocabulary to the modeling of present and future phenomena is, however, entirely absent from Bateson's work.

The question of temporality invited by Bateson's use of cybernetics as a purely diagnostic (rather than predictive) mode is helpful in locating the specific types of worldview that, when facing the possibility of cybernetic social theory, might successfully obviate the limitations and caveats detailed above. As has already been established, much of the earliest work in the cybernetic sciences is centered on the relationship between the present and the immediate future. Although the centrality of measurable activity that grounds the methods of Wiener, Rosenblueth, Bigelow, McCulloch, and Pitts, as well those of Bateson, clearly foreshadows the dividual and the perpetual recourse to dynamic flows of data and "results" in the control era, their version of cybernetics quite clearly lacks an economic dimension because as a mode of analysis it is limited to very short-term reflexive phenomena. Rosenblueth and Wiener, for example, presented work on the knee-jerk reflex at the second Macy conference and wrote with Bigelow about situations "so fast that it is not likely that nerve impulses would have time to arise at the retina, travel to the central nervous system and set up further impulses which would reach the muscles in time to modify the movement effectively."[32] To grasp *economic* behavior as functioning at the same level as instinctive reaction to stimuli is to uphold the classic assumption of liberal political economy—namely, that humans are intrinsically economic. To posit this economic being as (1) productive of homeostatic self-regulation across entire social systems and (2) requiring the construction and maintenance of a set of social, educational, and political (which is to say, epistemic) norms that valorize communicational exchange while rendering noncommunication aberrant is to add a "neo-" to this socioeconomic liberalism.

Indeed, a close examination of the arguments over the possibility of social modeling at the Macy conferences suggests that it was economics, rather than sociology or anthropology, that structured the critical break between those who cast doubt on social applications and those who actively pursued such applications. As Jean-Pierre Dupuy notes, participants in the debates over the social application of cybernetics at these conferences tended to polarize into two groups; those possessed of the "cast of mind peculiar to economists, which combines an obsessive concern for logical rigor with deliberately impoverished view of human relations," and those who "took advantage of every possibility to denigrate utility theory, which flattened all the components of desires and choice along a single dimension."[33] In the first of these groups, Dupuy lists John von Neumann, Leonard Savage,

Paul Lazarsfeld, and ("to a certain extent") Walter Pitts, and in the second group he includes Wiener, McCulloch, and Bateson.[34]

The contours of this basic opposition are illuminating if one seeks to track the intensification of the relationship between cybernetics and capital that developed through the second half of the twentieth century. The fundamental differences between two valences of cybernetics—one appearing as a stage in capital's expansion toward the informatic mode of real subsumption and one perhaps naively posited as incompatible with capitalist modes of production and social organization—can be illustrated through two focal points in the opposed perspectives that emerged in the cybernetics group: the first concerns the relationship between games and social activity in general, and the second is centered on the depth to which computational metaphors for the human can be extended.

Games as Worlds, Worlds as Games

Returning to the account of the antiaircraft predictor in the introduction to *Cybernetics*, it is notable that Wiener puts forward the "good poker player" as an example of a human actor that would fall outside of the predictive possibilities offered by cybernetic modeling. By contrast, in a note at the end of *Cybernetics* he writes enthusiastically on the possibility of constructing a chess-playing machine—an interest taken up by Shannon, who published an article titled "Programming a Computer to play Chess" in 1950 and who developed the first chess-playing computer (albeit one that had only enough processing power to play the endgame—thus its name, "Endgame"), as well as designing machines capable of playing Hex and Nim and solving mazes and Rubik cubes. For Wiener and Shannon, chess can be modeled and automated, but poker cannot. The reasons for this distinction can be found in the formal and material character of the games in question. Chess is a game of perfect information in which all players can see every element of the game at all times. It is, in Deleuze and Guattari's parlance, a game in which "pieces are coded," with "an internal nature and intrinsic properties from which their movements, situations, and confrontations derive."[35]

Poker, by contrast to chess, is a game of imperfect information in which central tactics include inferring an opponent's future actions and providing him or her with inputs (bluffing) in order to condition these actions. It is telling, then, that von Neumann was uninterested in chess but fascinated by the possibility of applying prediction and modeling to poker. In *The Ascent of Man*, Jacob Bronowski recalls a conversation with von Neumann in the early 1940s in which the latter stated, "Chess is not a game. Chess is a well-defined form of computation. You may not be able to work out the answers, but in theory there must be a solution. Now real games ... are not like that at all. Real life is not like that. Real life consists of bluffing, of little tactics of deception, of asking yourself what is the other man going to think I mean to do. And this is what games are about in my theory."[36] Although the neoliberal principle of social interaction as perpetual competition is writ large in this statement, equally telling is the implication that von Neumann's vision of the relationship between cybernetic thought and society, in distinction to Wiener's view, was directed toward finding ways to model and predict social situations of incomplete knowledge in service of maximal profit. It should be noted that von Neumann was not exactly saying that social interactions constitute a mode of computation in which every step is foreseeable. Rather, his vision of the world requires a process of conceptual coding in which concrete social existence is first rendered digital so that it can then be computed, and it is the *assumption of perpetual competition* that performs this prefatory digitization. Poker, in von Neumann's view, poses an engaging set of questions about computability because it stands in for a simplified microcosm of human interaction in a "real life" that, for him, consists of "bluffing," "deception," and the second-guessing and preemption of competing intentions over (one must infer) finite resources. Here it is quite straightforward to observe a more detailed picture of the division between the opposing poles in the early debates over cybernetics: as with their earliest experiments in antiaircraft prediction and statistical methods for communications engineering, Wiener and Shannon's engagement with games was limited to situations in which automation was possible only due to strict, already-existing, material limitations. Von Neumann's cybernetics constructs a worldview in which the contingent real of the social can be made the object of modeling and prediction by first being conceptualized in a way that reduces it to a minimal set of differential functions.

As von Neumann makes clear in his comment to Bronowski, the game-theoretical model of social interaction is directed toward rendering human behavior modelable, and this rendering is facilitated by limiting the conceptualization of this behavior with the assumption that it is always motivated by competition, rational calculation, and the maximization of gains.[37] This, it is worth noting here, is comparable to the way in which capital can view human life only as labor (and thus as interchangeable with machines, water, or air), according to Marx's description in the *Grundrisse*. This assumption, which forms the basic condition of possibility for game-theoretical politics and economics, was the subject of a number of strident critiques in the foundational years of the cybernetic sciences—critiques that predictably followed the division within the group of cybernetic theorists identified earlier. The character of these critiques is instructive in the attempt to clarify the connections between the socioeconomic valences of cybernetics and the logic of control.

The critiques of game theory that emerged in the 1940s can be divided into two central claims: game theory is premised on an impoverished vision of the human and of social groups that makes it inaccurate at best and politically dangerous at worst; and game theory, because of the centrality it grants to competitive economic motivations most clearly associated with free-market capitalism, is oriented toward maximizing accumulation for a few individuals or institutions and thus runs counter to the cybernetic goal of homeostasis. Von Neumann and Oskar Morgenstern actually preempt and dismiss the first of these criticisms in *Theory of Games and Economic Behavior* when they state, in relation to the prospect of developing a mathematical economics in general, that "the arguments often heard that because of the human element ... mathematics will find no application" in the analysis of economics "can all be dismissed as utterly mistaken."[38] This preemptive dismissal, however, did little to ward off the critiques of game theory put forward by Wiener and Bateson, among others. In the conclusion to *Cybernetics*, for example, Wiener writes that "von Neumann's picture of the player [and, by extension, every human] as a completely intelligent, completely ruthless person, is an abstraction and a perversion of the facts ... where knaves assemble, there will always be fools."[39] Wiener's objection can be understood as primarily scientific, rather than ethical: it is the abstraction and perversion of the "facts" of social life (i.e., the inaccurate definition of the objects of analysis) that makes accurate modeling of social behavior via game theory methodologically unsound.

Bateson broadens the implications of this critique of game theory from methodological feasibility to political danger in a 1952 letter to Wiener. Here Bateson writes that the deployment of game-theoretical approaches to social interactions, whether political (as in the case of the Cold War) or economic (as in the case of the burgeoning relationship between the RAND Corporation and the Cowles Commission), would serve only to "reinforce the players' acceptance of rules," "make it more and more difficult for the players to conceive that there might be other ways of meeting and dealing with each other," propagate a sense that "human nature is unchangeable," and produce a vision of "people and mammals" as robots who "completely lack humor and are totally unable to 'play' (in the sense in which the word applies to humans and puppies)."[40] The relationship between the vision of the von Neumannian robot and the reinforcing of particular behaviors expressed by Bateson is striking. As with the function of neoliberal governmentality, under the core tenets of the control episteme the subject must (and will) 'learn' how to act as an information processor (such as a game theorist) even as this form of sociality is presented as natural due to apparent homologies with the function of neurons, cells, and even the universe as a whole. Ultimately, for Bateson, the application of game theory to the conceptualization of social interaction (a practice that the RAND Corporation was actively developing at the time Bateson wrote to Wiener) "only propagates the theory by reinforcing the hostility of the diplomats, and in general forcing people to regard themselves and each other as von Neumannian robots."[41] The danger of the highly limited game-theoretical conceptualization of the human was for Bateson more than an impediment to its feasibility as a mode of prediction. His letter to Wiener suggests that he foresaw an epistemic function of game theory in which the "von Neumannian robot" would become the dominant vision of humans engaged in social interaction under global capital, and that this vision would in turn be productive of new and deleterious norms, concrete behaviors, and situations.

Bateson's response to game theory foregrounds a basic opposition between diagnosis and prediction in conceptualizations of how cybernetic methods might function within the social sciences. In "Bali: The Value System of a Steady State," Bateson uses game theory to argue that Balinese culture is premised neither on growth nor stasis, but rather on "continual nonprogressive change," emblematized by "ceremonial and artistic tasks which are not economically or competitively determined."[42] As noted earlier, cybernetic concepts in Bateson are directed not toward prediction but toward providing a vocabulary with which to analyze and compare already-existing social structures. By adopting this approach, Batson was able, for example, to compare the Iatmul with the Balinese on the grounds that the former is schismogenetic, characterized by "regenerative causal circuits or vicious circles" (positive feedback), whereas the latter is nonschismogenetic and characterized by negative feedback.[43]

Distinguishing diagnostic and predictive tendencies in social cybernetics is not intended as an absolution of Bateson (or of Wiener, who proposed similar diagnostic applications in his 1948 book and its 1950 follow-up *The Human Use of Human Beings*), especially considering that the diagnostic method, as will be shown later, is central to the cybernetic and system-dynamic management techniques that are emblematic of the structure and operation of post-Fordist business. The diagnostic application was itself the subject of stern critique in the 1940s, not least from the artist Maya

Deren, who saw Bateson present material pertaining to his work on Balinese culture in February and March 1947. Deren wrote in her notebooks
of her consternation over Bateson's use of cybernetic concepts, particularly
with regard to the use of a single technical vocabulary to express the differences between distinct societies such as the Iatmul and the Balinese.[44]
Echoing the methodological concerns of Wiener, Bigelow, McCuloch, and
Pitts detailed earlier, Deren argues that cybernetic methods should not be
applied even to the retrospective (i.e., non-predictive) analysis of social
structures because they fail to account for the passage of time: to claim that
societies are comparable in terms of universal concepts such as feedback
and homeostasis, she argues, is to suggest that these societies are formally
identical at all times and to ignore the existence of conscious (as opposed to
in-built, automatic) processes of correction and adaption as well as cultural
and individual memory.[45] It is possible to locate some fascinating parallels
between Deren's critiques of cybernetic social analysis and the directions
von Neumann pursued in his later work on computing machines and cellular automata. First, however, it is necessary to connect the methodological and predictive critiques detailed above to explicit debates around the
relationship between game theory and capitalist organization at the end of
the 1940s.

Bateson's concern with questions of growth and balance points to the
second major critique leveled at game theory—its role in a broader movement toward using a cybernetically oriented version of economics (specifically capitalist, free-market economics) to conceptualize and model
the social. Perhaps surprisingly, this critique is strongly voiced by Wiener
at certain moments in *Cybernetics*. At several points in the book, Wiener
clearly states that a truly homeostatic, communication-based model of society would be structurally incompatible with capitalist modes of production and thus cannot be realized while the latter systems exist. Toward the
end of the introduction to *Cybernetics*, for example, Wiener makes a clear
statement against (a somewhat limited understanding of) capitalism and,
in a remarkable turn of phrase, appears to suggest the necessity of revolution: "The answer [to the poverty that results from industrial automation],
of course, is to have a society based on human values other than buying
or selling. To arrive at this society we need a good deal of planning and
a good deal of struggle—which, if the best comes to the best, may be on
the plane of ideas, and otherwise—who knows?"[46] The tragedy of Wiener's
optimistic position on the prospective utopianism of cybernetics lies in his
mistaken location of commodity exchange, rather than the abstraction
and exploitation of activity as labor, as the cornerstone of capitalism. For

him, there is a basic incompatibility between a society governed by cybernetic principles and the primacy of capitalist accumulation, even though,
as this book argues, there already existed the nucleus of formal identification between the two in the nineteenth century, with the conceptual
frameworks set out by Wiener himself, von Neumann, and others appearing as a later stage in the widespread emergence of this identification under
the control episteme. The development and diffusion of this association,
which clearly contravenes major aspects of the pioneering work by Wiener,
Bigelow, Rosenblueth, McCulloch, Bateson, and Mead, among others, can
be clearly grasped through the emergence of computational metaphors for
human activity that represent the major endowment passed from cybernetics to the present political-economic orthodoxy.

The objections to game theory were not lost on von Neumann, although
his acknowledgment of them did not share in any of Wiener, Bateson, or
Deren's grounding ethical positions. Where Wiener, Bateson, and Deren,
among others, critiqued game theory for its impoverished vision of the
social and its fundamental assumption of competition as the baseline of
human interaction, von Neumann appeared to view these criticisms as
indicative of game theory's failure to incorporate complexity, communication, and self-correction (the same failure that Deren identified in the use of
game theory as an analytic model for social systems). For von Neumann, the
problem of the "von Neumannian robot" was not its cynical or reductive
view of the social human but the rigidity of its guiding metaphor—game-
theoretical robots can carry out preprogrammed behaviors automatically
but cannot adapt to new inputs by developing new behaviors. Rather than
discarding the core objective of game theory—the mathematical modeling
of dynamic systems, including social interactions—von Neumann retained
the basic principle that human behavior and thus social systems can be
subject to informatic modeling, but he also moved on to a wide-ranging
set of investigations around self-regulating and self-reproducing machines
(rather than preprogrammed and nonadaptive ones) as universal metaphors applicable across a range of complex systems. As Philip Mirowski
notes, von Neumann's "withdrawal from any further work on game theory
after 1944, with a few notable exceptions, dovetails with his turn towards
the theory of computation, brain science, and the theory of automata," the
last of which "stands as his most profound contribution to economics."[47]

Once again, this investigation into the contours of the debates between
the group of economically oriented cyberneticians (such as von Neumann)
and their critics (such as Bateson, Deren, and Wiener) is not an attempt to
paint the latter as heroes and the former as villains. As should be clear by

now, and as will be argued in greater detail in this chapter, elements of both groups' work have proven central to the mutation of capital in the transition to control societies. This comparison of the often opposed positions toward socioeconomic applications in the foundational years of the cybernetic sciences is instead intended to show that the integration of capital and cybernetics was far from an intended or even an obvious outcome, and that a complex series of arguments, translations, and mistranslations took place that made this integration appear not just desirable and possible but, paradoxically, both natural and necessarily bound to the internalization and normalization of particular conceptual structures.

Machinic Metaphors

Theories of cybernetics, in both celebratory and critical valences, are characterized by a clearly defined system of periodization in which certain technical systems are located as emblematic metaphors for a range of phenomena, from the human to the laws of physics in general, within successive historical epochs. This is apparent in Wiener's *Cybernetics*, where he plainly states that "if the seventeenth and early eighteenth centuries are the age of clocks, and the later eighteenth and early nineteenth centuries constitute the age of steam engines, the present time is the age of communication and control."[48] The centrality of cybernetics to Deleuze's control-society thesis is underscored by the fact that he retains this periodization theory in "Postscript." Here Deleuze makes the oft-cited announcement that "it's easy to set up a correspondence between any society and some kind of machine," before going on to imbricate Wiener's definition of successive technological eras with the extended Foucauldian periodization it is the purpose of the "Postscript" to develop, so that "the old sovereign societies" worked with "levers, pulleys, clocks," disciplinary societies with "thermodynamic machines," and control societies with "information technology and computers."[49] Another layer is added to the second and third stages of this periodization theory by the association of energetics and psychoanalysis in much cybernetics literature, as well as by the subsequent disavowal of this complex in favor of the adoption of computers as metaphors for the psychic apparatus. In a 1947 essay titled "The Fallacious Use of Quantitative Concepts in Dynamic Psychology," for example, Lawrence Kubie, the sole psychoanalyst among the first wave of cyberneticians, locates in the economic stage of Freudian psychoanalysis a "descriptive shortcut" of "strong and weak libidos, strong and weak ids, strong and weak superegos, strong and weak egos" based on the thermodynamic principle of "hypothetical changes of energy."[50] This association is repeated throughout a range of cybernetically inflected texts, notably in Wiener's tentative comparison of Freudian psychoanalysis to Gibbsian thermodynamics in *The Human Use of Human Beings*.[51]

The computer–human metaphor coalesces in debates surrounding cybernetics in the 1940s, but it is far from universal in its use and acceptance across the work of distinct practitioners. Once again, the difference in Wiener's and von Neumann's use of this metaphor is broadly indicative of the socioeconomic and political currents that shape its subsequent deployment as the epistemic ground of control-era capitalism. In this instance, it is time that motivates the differences between these two approaches to the computer as a conceptual analog for the human actor.

As discussed earlier, Wiener posits his work on computing machines and his study of electronic networks as critical precursors to the cybernetic approach to biological and technical systems. His identification of computing machines and network architectures as foundational to his collaborative work on the approach that would be named cybernetics is instructive in identifying a division in the conceptualization of the computer as metaphor and the uses it might be put to. As has already been established, for Wiener and his collaborators instrumental analogies between hardware computing and biological and social systems could be extended only if bounded by clear material limitations (for example, the split-second time frame of a neuron firing or an instinctive [if purposeful] reaction such as a fighter pilot's reaction to an incoming projectile), and for these early cyberneticians such limitations provide a clear reason why social phenomena such as the valorization of surplus labor in an expanded field (such as across all cognition and social interaction) fall outside of the realm of possibility for cybernetic modeling. It thus becomes possible to note a technical-temporal character to Wiener's optimistic claim that cybernetics and capitalism are basically incompatible. In a 1960 article titled "Some Moral and Technical Consequences of Automation," Wiener writes that "one of the chief causes of the disastrous consequences in the use of the learning machine is that man and machine operate on two different time scales, so that the machine is much faster than man and the two do not gear together without some serious difficulties. Problems of the same sort arise whenever two control operators on very different time scales act together, irrespective of which system is faster and which system is slower."[52] Although Wiener's primary investment here is in practical matters of human–machine interaction, the implications for the cybernetic human–computer metaphor are significant: in Wiener's cybernetics, it is only with "serious difficulties" that the temporal scale of the machine can be "geared together" with that of the human. This definition is problematic for several reasons, but the most pressing is the way in which it supposes that "time scales" are unequivocally synchronic, existing outside of historical conditions and thus not subject to mutation by historical forces.

Under the temporal incompatibility between human and computer Wiener sets out, some universal and immutable "human time" can meet "computer time" only with "disastrous consequences." Wiener's approach thus precludes the question of this "computer time" affecting a transformation on at least one scale of "human time," just as the mechanical clock did in the passage from sovereign to disciplinary societies. The possibility that this "computer time" is in fact closer to the temporality implied by the abstract logic of capital (as distinct from the specific production of linear time necessary for the concrete conditions of production Marx encountered) than to the blunter, more rigid time imposed by the linear clock is equally impossible to determine under the schema Wiener sets out. This impossibility, it must be observed, is consistent with Wiener's faulty equation of capitalist social relations with the exchange of commodities rather than with the social relations of exploited labor time. His conceptualization of humans and of capital thus presents a clear tension with the broader cybernetic principle that humans and machines are conceptually interchangeable. The resolution of this tension is central to von Neumann's foundational work on computing machines, and it is this resolution that allows the computer metaphor to assume a central role in conceptualizations of psychology, management, and economy, where it had previously been limited to very specific areas of physiology.

In systematically working through the technical considerations necessary for the building of stored-program electronic digital computers (namely, the Electronic Discrete Variable Automatic Computer, EDVAC), as recounted in "First Draft of a Report on the EDVAC" of 1945, von Neumann considers two possibilities for synchronizing the actions of a machine's various parts: the first lies in the production of an "autonomous" system in which "the successive reaction times of its elements" determine the rhythms of the machine; the alternative consists of a system in which the timing of the various hardware elements are "impressed by a fixed clock, which provides certain stimuli that are necessary for its functioning at definite periodically recurring moments."[53] The general concept of the electronic digital computer, in von Neumann's theorization, does not carry with it a single, inherent temporal logic: a computing machine might equally be built on either an internal, flexible temporality or an external, rigid one. The "clock impressed timing," von Neumann later states, emerges as preferable only "if reliance is to be placed on synchronisms of several distinct sequences of operations performed simultaneously by the device."[54] Synchronicity of spatially separate components, then, would necessitate the use of the linear, periodic clock in computation, just as the necessary synchronization of time across distant geographical spaces (to prevent things like railway accidents, of course, but also to facilitate trade) recurs throughout the history of capital's spatial expansion through colonization.

Von Neumann returns to the temporality of computation in the next section of "First Draft" (4.2). Here he examines the principle of timing logic gates in order to draw a basic identification between the stored-program computer and human cognition, the latter in the idealized form of McCulloch and Pitts's simplified model of neural function in biological life. As von Neumann writes, "It is worth mentioning that the neurons of higher animals above sense are definitely elements in the above sense [of binary switches]. They have an all-or-none character, that is two states: Quiescent and excited."[55] This passage, which takes the model of the brain as an inspiration for computer-hardware design rather than the reverse, must be seen, along with Babbage's intertwined theories of factory machines, computation, the universe as a whole, and the division of labor, as a major moment of conceptualization in the mutation of capitalism toward an informatic mode. Once again it is notable that in establishing this identification between brain of the "higher animal" and the computing machine, von Neumann concerns himself principally with time. The neuron of the "higher animal," von Neumann observes, "has a definite reaction time, between the reception of a stimulus and the emission of the stimuli caused by it," and so falls into the category of the asynchronous (that is, internally and unevenly timed) computer.[56] Between the lines of this statement, one finds a reformulation of the problematic comparison between human time and machine time that Wiener sets out: the "asynchronous" cognitive function of the "higher animal" is incompatible with "synchronous" time of the type regulated by the oscillations of some mechanical, electrical, or electromechanical element. In von Neumann's comparison, the time of human cognition is radically different from that of the linear clock, although this distinction results from a difference in regularity rather than from the difference in speed that Wiener posits as the basic incommensurability between human and computer.

Predictably enough, von Neumann—the figurehead of "cyborg economics," as Mirowski has it—shows little interest in allowing the temporal incommensurability between computer time and human time to serve as an impediment to their conflation by way of cybernetic logic. A significant part of the "First Draft" is thus engaged with the potential of computer hardware to overcome this incommensurability. Sections 5 and 6 are concerned with increasing the computer's efficiency and speed in performing arithmetical operations or processes, and across the two sections one can observe the resolution first of Wiener's difference of time scales and second of von Neumann's more fundamental synchronicity difference. In section 5, "Principles Governing Arithmetical Operations," von Neumann considers two opposed methods of extracting maximum speed from his hypothetical hardware: "*telescoping operations*" ("carrying out simultaneously as many as possible") and making the device as simple as possible ("never performing two operations simultaneously").[57] Von Neumann quickly discards the first of these methods (telescoping operations), despite its being the "logical solution" and the one "being used in all existing devices," and puts forth the second method (simplifying the device) as preferable. The temporality of all subsequent computing machines that use the von Neumann architecture is thus established as an anthropomorphization of the machine and a mechanization of the human, a system of equalizing human and machine temporality through the reproduction of the former at a smaller scale.

The discontinuity artificially bridged by von Neumann's foundational abstraction can be clearly evidenced by reference to a more recent textbook on hardware–software interfaces:

Although as computer users we care about time, when we examine the details of the machine it's convenient to think about performance in other metrics. In particular, computer designers may want to think about a machine by using a measure that relates to how fast the hardware can perform basic functions. Almost all computers are constructed using a clock that runs at a constant rate and determines when events take place in the hardware. These discrete time intervals are called *clock cycles* (or ticks, clock ticks, clock periods, clocks, cycles). Designers refer to the length of a *clock period* as the time for a complete *clock cycle* (e.g. 2 nanoseconds, or 2ns) and as the *clock rate* (e.g. 500 megahertz, or 500mhz), which is the inverse of the clock period.[58]

Here the artificial collapsing of human and computer at the level of time is laid bare: we, as computer users, "care about our time," but this consciously apprehended time is apparently not related to the micromeasures of time metronomically demarcated by our hardware. Computer clock cycles, in Patterson and Hennessy's telling, somehow represent both time and some "other [presumably nontemporal] metric." The solution to this tension, Patterson and Hennessy go on to argue, is to "formalize the relationship between the clock cycles of the hardware designer and the seconds of the computer user."[59] The status of computation within the digital logic of capital becomes clear in this abstraction: computing machines serve as both tool and metaphor (just as their mechanical and thermodynamic predecessors did), and the specific result of this correlation is that although computer time remains imperceptible on a human scale due to hardware and software design, there is an invisible impression of computer time on the human user. The very specific frustration of having to wait, clicking and hitting keys with no discernible effect, while a computer resolves some incommensurability between queued processes and available CPU cycles attests to the power of this illusion and its centrality to the social temporality of computing.

If von Neumann's work on computing machines provides the cybernetic reconceptualization of the human with a temporal axis, his later work on the general theory of self-reproducing automata extends this formulation into a second dimension, that of space. Mirowski views von Neumann's final decade of work (from the late 1940s to his death in 1957) in terms of a progressive displacement of game theory—which nonetheless retained a significant influence on think tanks such as the RAND Corporation—by the theory of automata. As Mirowski writes, "In the transition between game-theoretic rationality and automata some aspects of the formalism are preserved, while others are amended or summarily abandoned. Most obviously, where game theory tended to suppress formal treatment of communication and the role of information, the theory of automata elevates them to pride of place. The very idea of a roster of possible moves in the game is transmuted into an enumeration of machine states. Strategies, a foundational concept of game theory, now become internally stored programs."[60] This overview of von Neumann's work on automata foregrounds two central elements: (1) that the theory does not do away with the ideal of projective social modeling that underpins game theory but rather develops its fundamental aims by replacing the concept of the player with that of a cybernetic machine capable of self-regulation through communication and feedback even when unforeseen inputs arise (a development that, intentionally or not, responds to the formal but not the political critiques of game theory set out by Wiener, Bateson, and Deren, among others); and (2) that the imbrications of computing machine and human that run through "First Draft of a Report on the EDVAC" are developed into the central metaphor in a universal theory of representation and predictive modeling. Where the theory of computing elaborated in the "First Draft of a Report on the EDVAC" works though a temporal equalization of human and machine "thought," von Neumann's theory of automata is premised on establishing a more fundamental identification between the two that does away with the need for such explicit conceptual and technological equalization.

The functional erasure of the gap between the human and the mathematically modeled automaton was the first concern addressed when von Neumann presented his theory of automata to the Hixon Symposium on September 20, 1948.[61] In his introductory remarks, von Neumann both identifies the incommensurability of natural and artificial systems and states that this incommensurability does not invalidate the usefulness of analogies between the two for analysis and planning: "Natural organisms are, as a rule, much more complicated and subtle, and therefore much less well understood in detail, than are artificial automata. Nevertheless, some regularities which we observe in the organization of the former may be quite instructive in our thinking and planning of the latter; and conversely, a good deal of our experiences and difficulties with our artificial automata can be to some extent projected on our interpretations of natural organisms."[62] What this statement makes clear is that von Neumann's late work does not exactly argue that humans and machines are the same. Rather, the method pursued in this work attempts to establish the differences between the two as trivial for procedures of mathematical representation and planning. In these opening remarks, it is already possible to note the progression from the tentative analogy between human and computing machine developed in "First Draft of a Report on the EDVAC" to a project that takes a fundamental homology between the two as a basis for approaching more complex problems of self-organization.[63]

The remainder of "The General and Logical Theory of Automata" is concerned with working through problems of analogy between biological central nervous systems and technological computing machines. Here von Neumann argues that the living organism is of a mixed character, with the digital neurons theorized by McCulloch and Pitts existing in often complex chains and feedback loops with analog phenomena such as "the general chemical composition of the blood stream or of other humoral media."[64] The posthumously published book *Theory of Self-Reproducing Automata*, a series of talks and drafts from 1949 and 1952–1953 collected and annotated by Arthur W. Burks, sets out an expanded discussion of the concerns initially developed in "General and Logical Theory of Automata." The project detailed across these works represents a sustained, technical discussion of the idealized human–computer metaphor that must be seen as fundamental to the applicability of cybernetic, predictive modeling to society as well as to computing machines and individual neural nets.

Although there is no explicit discussion of social systems in either "General and Logical Theory of Automata" or *Theory of Self-Reproducing Automata*, the entire project, with its detailed elaboration of a way in which the difference between humans and computing machines is rendered trivial from the point of view of formal logic, can be read as a working-around of the shortfalls that critics identified in game theory and in the early discussion of social cybernetics. The worldview that both produces and is produced by this approach becomes clear when Burks, who worked on the EDVAC project with von Neumann in 1945, points out in his annotations to *Theory of Self-Reproducing Automata* that there is a "striking analogy" between automata theory and game theory because "economic systems are natural," whereas "games are artificial," so "the theory of games contains the mathematics common to both economic systems and games, just as automata theory contains the mathematics common to both natural and artificial automata."[65] The principal concerns discussed in the talks that comprise *Theory of Self-Reproducing Automata* center on the ways in which living organisms' capacity to continue functioning in the presence of errors might be reproduced in artificial automata and on the possibility of constructing artificial machines with the capacity to self-reproduce in ways that can generate novel outcomes dependent on feedback. These two principles—the conceptualization of the human as an ideal distributed network and the artificial production of novelty out of rules—clearly prefigure the tenets of control with its valorization of communication and the seemingly contradictory centrality it attributes to both algorithmic constraint and cultural innovation. Von Neumann's work on automata theory might not explicitly contain a theory of society, but it does represent a critical step in the conceptualization of cybernetic logic, initially constrained from social applications by the idealized character of digital neurons and the materiality of metal as much as by ethical concerns, into a world picture that is optimal for the real subsumption of sociality by capital.[66]

The preceding paragraphs, although focused on the tension between different conceptualizations of cybernetics as they play out in the work of Wiener and von Neumann, are not intended to apportion these two mathematicians with some dubious privilege within the recent history of capitalism. They are not intended to paint the two as privileged arbiters of particular worldviews tussling to shape history through theoretical and applied mathematics. Such an argument would be absurd. Rather, the specific analysis of their work given here is intended to foreground the contours of a broader epistemic shift centered on the prospective notions of cognition and society as digitally representable and on questions about whether the projective modeling of social phenomena would be possible and desirable, as well as to highlight the necessary conceptualizations of the individual human actor and of society required to answer these questions in the positive. In short, Wiener and von Neumann are here used to exemplify particular, polarized worldviews shared by many others either directly working on cybernetics research or working in apparently distinct fields of scientific, economic, and political inquiry and practice. Having worked through the complexities and tensions through which these reconceptualizations of the subject and social interaction are shaped, one can now turn to some of the ways in which, following the initial tensions examined earlier, the imbrications of cybernetics and capital developed from the late 1940s on. By tracking some of the threads that connect cybernetic logic with the socioeconomic formations of the present, one might be able to establish a foundation for inquiry into the *cultural* valences of control that attest to the breadth and depth of its penetration.

Cybernetic Society

In April 1948, the philosopher F. S. C. Northrop, an attendee at each of the first eight Macy conferences, published an article in *Science* proposing that, contra Wiener, cybernetic methods and terminology make the formulation of a "normative" theory in the social sciences "both possible and significant."[67] Throughout this article, Northrop proposes that ideas constitute a social correlate to the environmental factors that concerned the physiological studies from which the core cybernetic methods developed—a move that places his approach far closer to von Neumann's language-based, general automata than to the specific, tightly restricted studies presented by Bremer, Gerard, Harrower, William Livingston, McCulloch, Pitts, Rosenblueth, Gerhardt von Bonin, Wiener, and others at the first three Macy conferences.[68] The majority of Northrop's paper is taken up with mapping the McCulloch–Pitts neuronal network and the Rosenblueth–Wiener–Bigelow concept of purposefulness onto social structure. The outcome of this mapping is remarkable for its anticipation of the universality of distributed networks and of pop-cultural (as well as economic-theoretical) neuroscience in the control era:

Cultural factors are related to biological factors in social institutions by the biologically-defined purposeful behavior of human neurological systems containing negative feedback mechanisms and the normative social theory defined in terms of the universals which are the epistemic correlates of trains of impulses in neural nets that are reverberating circuits. Because overt behavior can be tripped by impulses from reverberating circuits whose activity conforms to universals, as well as by impulses coming immediately from an external particular event, the behavior of men can be, and is, causally determined by embodiments of ideas as well as by particular environmental facts. And since the brains of men in early so-called primitive societies are provided with reverberating circuits, just as are the brains of men in so-called modern societies, it follows, though the specific universals may be different, that normative social philosophies will be significant in any culture.[69]

The ease with which events, environmental facts, and ideas are rendered equal, the collapsing of the cultural and the biological via the network form, the erasure of any distinction between "so-called primitive" and "so-called modern" societies through the universality of the reverberating circuit, and the suggestion that behavior can be causally determined by ideas as inputs all emphasize the role of cybernetic logic in the formation of systems of thought amenable to the principles that constitute the control episteme. Northrop's conclusions, which extend the (once again) specific and limited principles of hypnotism discussed in the 1942 cerebral inhibition meeting that preceded the Macy conferences into a general theory of programmability based on communicational inputs, provide a transhistorical model of the social and cultural actor as information processor that is identical to the kinds of conceptualization that are essential for notions such as information work and affective labor to function.

Beyond the contributions of Macy conference participants such as such as Bateson and Northrop, the diffusion of cybernetics into social theory can be observed in the work of Claude Lévi-Strauss. In an essay first presented at the Conference of Americanists in 1949, Lévi-Strauss responds to Wiener's negative appraisal of the applicability of cybernetic methods in the social sciences by claiming that "there is ... at least one area of the social sciences where Wiener's objections do not seem to be applicable, where the conditions which he sets as a requirement for a valid mathematical study seem to be rigorously met. This is the field of language, when studied in the light of structural linguistics, with particular reference to phonemics."[70] Following this statement, in which the symbolic overcoding that *langue* effects on *parole* emerges as analogous to the digital reduction of social reality to discrete, qualitatively compressed units, Lévi-Strauss moves on to ask the following questions: "Is it possible to effect a similar reduction in the analysis of other forms of social phenomena? If so, would this analysis lead to the same result? And if the answer to this last question is in the affirmative, can we conclude that all forms of social life are substantially of the same nature—that is, do they consist of systems of behavior that represent the projection, on the level of conscious and socialized thought, of universal laws which regulate the unconscious activities of the mind?"[71] In formulating these questions, Lévi-Strauss demonstrates the ease with which the basic principles of cybernetic approaches to biological and technological regulation were extended beyond these fields into wider and wider theories of behavior and society starting in the late 1940s. The result of answering these questions in the positive, as Lévi-Strauss does, is a vision of sociality that is essentially similar to that posted in Northrop's 1948 paper. Both of these accounts rest on a metaphorized version of cybernetics, a vision of the world as a network of oscillating probabilities in which the structuralist concept of language as a system of discrete units can be scaled up to model "forms of social phenomena."

It is fascinating to note, while tracing the emerging relationship between cybernetic social logic and the function of capital, that Lévi-Strauss's first area of consideration in thinking about the possibility of modeling social phenomena is "the question of why a particular style [of woman's dress] pleases us or falls into disuse."[72] After an initial process of "measuring some basic relations between the various elements of costume," Lévi-Strauss concludes, "the relationship thus obtained can be expressed in terms of mathematical functions, whose values, calculated at a given moment, make prediction possible."[73] The possibility of predicting trends by a process of symbolic representation and mathematization depicted by Lévi-Strauss here is remarkably similar to Lazzarato's definition of immaterial labor as comprising "the kinds of activities involved in defining and fixing cultural and artistic standards, fashions, tastes, consumer norms, and, more strategically, public opinion."[74]

In addition to unintentionally foregrounding the basic connections between structuralist linguistics, cybernetic logic, and the informatic stage of capitalism, Lévi-Strauss's essay is notable for containing an attempt, albeit an unconvincing one, to foreclose the cultural-political critiques necessitated by the prospective symbolic formulation and management of the social. Lévi-Strauss's response to "charges of 'anti-feminism'" leveled against cybernetically inflected structuralist approaches—in which "women are referred to as objects" in accounts of marriage regulations and kinship, for example—is grimly fascinating when placed in parallel with the very real socioeconomic and global implications of control-era capitalism with its digital techniques of capture, definition, and exclusion.[75] In light of the material and symbolic violence effected by the various apparatuses of control, Lévi-Strauss's insistence that, regardless of the analytic uses of symbolic representation and modeling, actual social individuals will never lose their "character of value, to become reduced to pure signs" in any materially significant way today reads either as a spectacular failure of historical imagination or as an image of the cunning of history in full effect.[76]

If Lévi-Strauss and Northrop (as well as Batson) illustrate some of the vectors along which cybernetic methods became generalized through social theories in the late 1940s, more fully developed tendencies toward the real subsumption of sociality and affect that is of such central concern to Lazzarato, Virno, and Hardt and Negri, among others, can be observed in a number of projects developed after encounters with cybernetic logic in the 1950s. Perhaps the most dramatic example can be found in a paper titled *Social Interaction*, prepared for the RAND Corporation by Robert F. Bales in 1954. Mirowski notes that the strength of the commonplace association between RAND and Cold War policy poses an "unfortunate" impediment to grasping RAND's wider socioeconomic significance. Not only was RAND the "primary intellectual influence upon the Cowles Commission in the 1950s, which is tantamount to saying RAND was the inspiration for much of the advanced mathematical formalization of the neoclassical orthodoxy in the immediate postwar period," Mirowski argues, but it was also "the incubator for cyborgs inclined to venture out into the worlds of management, the military, and the social sciences."[77] Bales's report proposes a "systematic procedure for analyzing social interaction" in situations where the analyst must "gain information from the other person in a social relationship without asking directly."[78] Here one can immediately note the imperative of forecasting with incomplete information that motivates cybernetic practices from Wiener and Bigelow's antiaircraft predictor, Shannon's communications theory, and von Neumann and Morgenstern's game theory, now applied to social interaction in general—albeit a version of the latter that, Bales posits, is confined to a symbolic formulation of language, being "largely made up of the talking that people do when they get together."[79] Although Bales's paper does not mention cybernetics by name, the fundamental principles of that approach are quite evident in the method and worldview the paper espouses.

Bales's approach is grounded in a generalization of all activity as communication and an erasure of the distinction between humans and computing machines, so that "study of social interaction on the face-to-face level assumes a broad significance when one recognizes that what goes on in a small decision-making group is a microcosmic prototype of the processes and problems that characterize a wide variety of communication and control networks, both human and electronic."[80] This logical grounding is directly compared to an "air defense network—a more or less typical large scale communication and control system."[81] Bales's method, universal enough to account for both missile systems and conversations, allows for the creation of a typology of interaction units that can be used to notate and analyze all social interactions. Bales describes a study in which subjects were asked to "discuss a human relations problem of the sort typically faced by an administrator in his organization," and the process of observation and notation set out is quite remarkable:

As the subjects begin their discussion the interaction observers begin forty minutes of intensive work. Each observer has on the bench in front of him a small machine called an interaction recorder, which presents a paper tape onto which he can write. The tape moves horizontally under a narrow opening in the top face of the recorder at a constant speed. At the left side of the opening is a list of categories for classifying the behavior observed.

The observer has already memorized an identification number for each sub-ject. When a given member speaks, the observer decides rapidly what classification should be given to the act, finds the place on the paper tape opposite the correct category, and writes down first the identification number of the person speaking, followed by the identification number of the person spoken to. The unit classified as an act is essentially a single simple sentence, or a non-verbal equivalent, such as a nod or a frown. Acts ordinarily occur at a rate of fifteen to twenty per minute. The paper tape keeps a record of the acts in proper time sequences.[82]

Two things stand out here. First, the definition of the "unit classified as an act" consisting of a "single sentence" or "nonverbal equivalent" calls to mind the encoding of social phenomena as discrete symbols that Lévi-Strauss proposes, which is to say, a representational extension of the digitalizing logic of capital to sociality and communication described in chapter 1. These units, Bales later notes, represent "the end product of a long series of attempts to construct a satisfactory typology" so that "observ-ers can reach a definitive classification of any given act by at most four binary choices in a logical tree."[83] Through the social arrangement imaged here, social interaction becomes representable and modelable as a system of seven "component acts" derived from the complex extension of compu-tational metaphors. This is a far more direct description of the capture of humans' cognitive and/or linguistic capacities than is found anywhere in Lazzarato, Virno, Negri, or their ilk. Second, the description of the interac-tion recorder, with its continuously scrolling paper tape upon which sym-bols are inscribed, calls to mind the tape of Turing's universal machine or von Neumann's cellular automata, but in this instance the data written to the tape concern human sociality, the components that write and process these data are human observers, and the "computer" the tape feeds into is a programmatic model of formalized human sociality. The world picture produced by these elements of Bales's study—in which both the observed subjects and the total group consisting of both subjects and observers appear as computing machines, and the social world as totality appears as a program—is emblematic of what this book defines as the cultural logic of control.

Given the explicit orientation of these applications of cybernetic logic to sociality by Lévi-Strauss and Bales, it should come as no surprise that the universalized form of cybernetic logic begins to appear in mainstream economic and management theories after the late 1940s. It is unsurprising because, as Dupuy notes, the basic principle of imagining the economic actor as an ideally rational data processor is fundamental to neoclassical economics as it has existed at least since the turn of the twentieth century. As the discussions of Babbage and Hollerith in chapter 1 ought to have made clear, the principle of digitizing the social for the purposes of economism and governance did not begin with cybernetics. Rather, cybernetics provided this principle with an epistemic grounding in the form of the digital-computational concept of the human actor that coincides with an intensification in the movement of capital beyond the fixed time-spaces of the factory and the office and into the social field as a whole. As Dupuy puts it, "Because the spirit that animates cybernetic modeling is profoundly similar to that of the modeling employed in mathematical economics, it comes as no surprise that the various disciplines issuing from the first type of modeling (systems theory, operations research, optimal control theory, decision theory, and so forth) should have provided the second with some of its chief tools."[84] Both Dupuy and Mirowksi give extensive accounts of the ways in which cybernetics shaped economic discourses in the second half of the twentieth century, and space does not permit a thorough reproduction of these arguments. Equally, there is insufficient space here to recount the many direct, causal relationships between cybernetics and finance, such as the development of computer trading systems like the Bloomberg console, the emergence of algorithmic trading and stochastic forecasting methods, the theorization of linear programming under von Neumann's consultation at the Cowles Foundation, and the centrality of game theory to the RAND Corporation's many activities.[85] Rather, this section examines particular examples that emphasize the shifting conditions of knowledge undergirding conceptualizations of the human actor in economic theory after cybernetics.

The central concept around which the melding of cybernetics and economic theory took shape is that of complexity. The foundations of cybernetic methods in Bigelow and Wiener's antiaircraft predictor and von

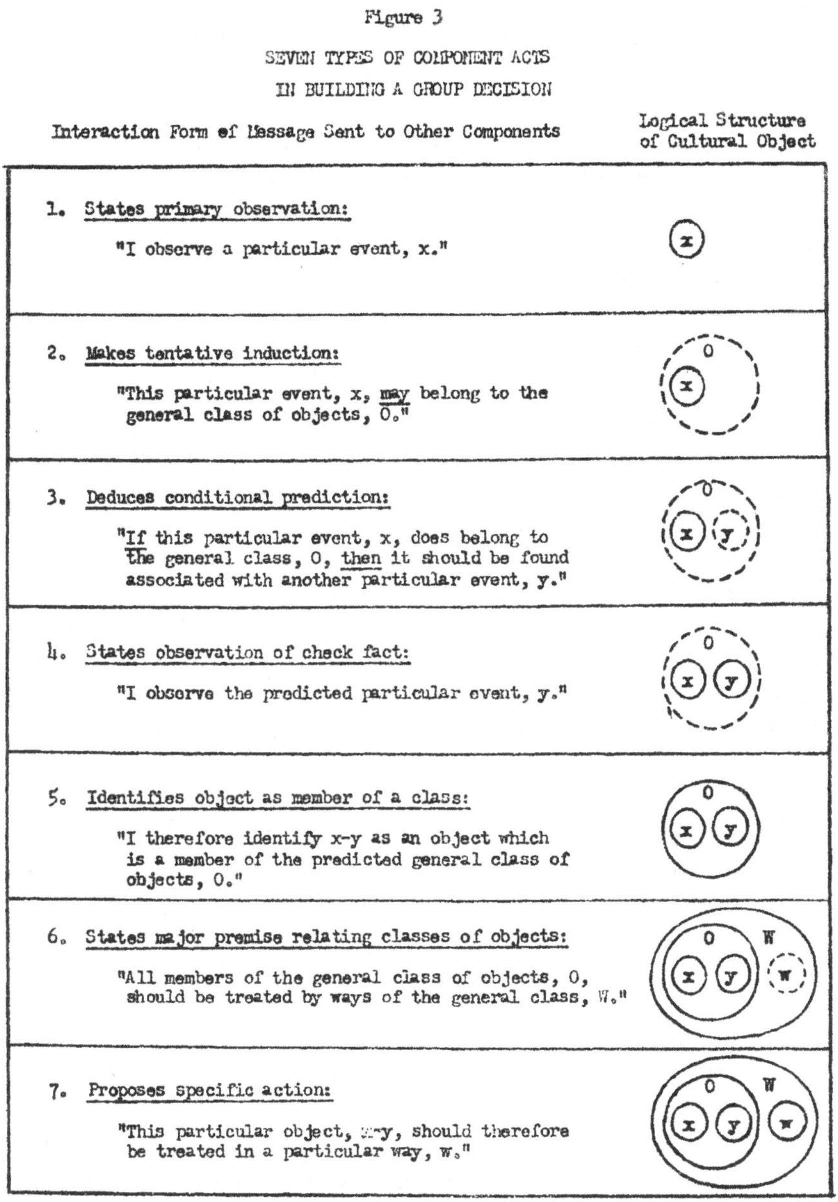

Figure 3

SEVEN TYPES OF COMPONENT ACTS

IN BUILDING A GROUP DECISION

Interaction Form of Message Sent to Other Components	Logical Structure of Cultural Object

1. States primary observation:

   "I observe a particular event, x."

2. Makes tentative induction:

   "This particular event, x, may belong to the general class of objects, O."

3. Deduces conditional prediction:

   "If this particular event, x, does belong to the general class, O, then it should be found associated with another particular event, y."

4. States observation of check fact:

   "I observe the predicted particular event, y."

5. Identifies object as member of a class:

   "I therefore identify x-y as an object which is a member of the predicted general class of objects, O."

6. States major premise relating classes of objects:

   "All members of the general class of objects, O, should be treated by ways of the general class, W."

7. Proposes specific action:

   "This particular object, x-y, should therefore be treated in a particular way, w."

Figure 2.1
Seven types of component acts in building a group decision. *Source*: Robert F. Bales, "Social Interaction," RAND Paper P-587, second draft, RAND Corporation, December 14, 1954. © RAND Corporation, 1971. Reprinted with permission.

Neumann and Morgenstern's game theory can be seen as the development, at very different scales, of systems through which to manage complex situations in which no human actor can hold all of the information necessary to assume such control through other means. As such, these methods bear an uncanny resemblance to the neoliberal principle that even collective human intervention into the economy cannot produce stability and growth because of the limits on human knowledge and that therefore only markets, carefully supported by governmental structures that do not regulate but only prevent impediments to flows of information and capital, can be trusted to assure "optimal" outcomes.

This affinity between the cybernetic view of complex systems and the principles of neoliberal political economy is crystallized in a letter from W. Ross Ashby published in the February 24, 1945, issue of *Nature*. In this letter, Ashby—a psychiatrist whose homeostat, an early example of a self-regulating machine developed between 1946 and 1948 and demonstrated at the ninth Macy conference in 1952, did much to illustrate the fundamental logic of cybernetics as a set of principles that function across biological and mechanical systems—uses cybernetic terminology to argue against state intervention into economic matters. Using the concepts of equilibrium and "vicious circles" (positive-feedback loops), he argues that "the introduction of governmental controls has led to many matters being dealt with by an order of *fixing* some quantity, price, or other variable where a *laissez-faire* system would have allowed them to find their own levels." The purpose of his letter, he goes on to state, is to "point out the danger that in any dynamic system the fixing of one variable may render the rest unstable." According to cybernetic principles, Ashby concludes, "(1) To fix a sociological or economic variable by order carries some danger of rendering the system, or parts of it, unstable (the latter being shown in the subsequent development of various 'vicious circles'). (2) The type of variable more particularly dangerous from this point of view is one which, under free conditions, changes value at high speed, and by these quick changes affecting the other variables, fluctuates only through a narrow range."[86] Such is the universality of cybernetic logic after its original material and methodological boundaries have been eroded. The only concern for the management of any dynamic system, be it biological, mechanical, or social, is the stability of certain key variables. It is telling, then, that seven pages before the draft of this letter appears in Ashby's diary, a short note is given demonstrating the "obvious" reason why civil servants should not show "spontaneous initiative" in introducing changes and reforms but should "wait until some variable becomes excessive" lest they "upset the stable system."[87] The question of whether this systemic stability would be evenly distributed remains unaddressed, as does the question of what constitutes an "excessive" variable. Ashby does, however, state that in building a stable society it is "important that all effects transmitted from one person to another should be graduated" rather than placed on a command/forbid dichotomy,[88] revealing the kind of self-regulation fundamentalism vis-à-vis social management that implies a less than optimal outcome for those who fall on the wrong side of a system's filters or thresholds.

The conceptual compatibility between neoliberal economic theory and cybernetic discourses revealed in Ashby's letter to *Nature* is further exemplified in the ways in which the latter intersect with the work of Friedrich Hayek. As both Dupuy and Mirowski point out, Hayek was preempting cybernetic discussions of information and modeling in arguments over economic theory from the 1930s on, and he sought out work in cybernetics that supported and expanded his fundamental approach.[89] In describing the period in which he was writing his psychological work *The Sensory Order*, Hayek recalls meeting von Neumann in the mid-1940s and discovering to his "amazement and delight" that von Neumann "immediately understood" what he was attempting with his work on psychology and "said that he was working on the same problem from the same angle." Following this comment, however, Hayek admits that although the later publication of von Neumann's work on automata showed him that he and von Neumann had been "thinking along similar lines," he found that he couldn't "profit" from it because it was "too abstract."[90] Because of this failure to fully grasp von Neumann's work, as well as a number of barbs aimed at Hayek by Morgenstern, Mirowski suggests that Hayek be regarded as "someone who filtered various cyborg themes into economics at second- and third-hand, motivated to search them out by his prior commitment to the metaphor of the market as a powerful information processor."[91]

To take Hayek's limited grasp of automata theory as a sign that the relationship between cybernetics and economics is merely circumstantial, however, would be to miss the fundamental character of the relationship. The version of cybernetics that is applicable to social modeling and that configures the social actor as an information processor is not the same as the version developed in Wiener, Bigelow, and Rosenblueth's work on purpose and teleology or in McCulloch and Pitts's protoneuroscience, Shannon's communication theory, or even Bateson and Mead's analytical deployment. Each of these developments provides the cybernetic mutation of capital with essential conditions of knowledge and possibility, but the version of cybernetics that might serve as an invitation to model and control social phenomena, whether economic, governmental, or military, results from the mutation of these metaphor-ready developments through specific historical forces—the same forces that shaped and elevated the market fundamentalism of which Hayekian theory is emblematic and that today reach beyond economic theory into a multitude of applications, from financial services to public health care and education.

In other words, the technicalities and mathematical principles found in the cybernetic sciences (including von Neumann's work) become eroded in the movement to social applications, leaving a version of the human–computer metaphor stripped of its specificities and caveats but retaining a historical momentum and sheen of epistemological justification from the technoscientific inquiries that produced it. Mirowski notes that "wave upon wave of computer metaphors keep welling up out of cybernetics, operations research, computer science, artificial intelligence, cognitive science, software engineering, and artificial life and washing up and over the economics profession with varying periodicities, amplitudes, and phase shifts,"[92] but he stops short of proposing that these metaphors continue to wash up alongside new computational methods for economic analysis and practice because they represent a precondition for the emergence of the informatic mode of capitalism rather than because they are rooted in any materially specific engagement with computing. With these informatic concepts of the social actor and the social group, which facilitate a superimposition of market and sociality on nominally objective and technologically timely grounds, came new modes of management and workplace organization that, along with the real subsumption of sociality, affect, identity, and so forth, represent the industrial logic of control. An analysis of these modes illustrates some of the ways in which the abstractions of cybernetic logic would be brought to bear on concrete social situations, with implications that reach beyond direct uses of communications technology.

Cybernetic Management

The social implications of cybernetic logic in the emerging period of control can be observed in the ways in which this logic has been turned toward the organization of labor. The work of Anthony Stafford Beer represents the earliest direct application of cybernetic principles to industrial management, and foregrounds both the process through which the former became fused with capitalist social organization and the fundamental identification between a cybernetic concept of management and the new and emerging modes of production characteristic of the developed forms of control society that are visible today.

From the 1950s until his death in 2002, Beer developed his practice of cybernetic management across a number of published texts and lectures as well as through practical work at United Steel, the International Publishing Company (now IPC Media), and (to foreground a quite different application) the socialist Project Cybersyn in Chile.[93] Each of these projects was premised on the universality of communication, feedback, and self-regulation across machines, bodies, brains, and economic systems. Beer's management cybernetics proposes a mode of organizing industry that is flexible and adaptive, allowing a given operation to respond to fluctuations in the "environment" in which it operates just as a biological organism responds, for example, to changes in temperature. Contra Wiener's insistence that "buying and selling" are antithetical to the equilibrium of social systems, Beer's 1959 model of industrial organization proposes that a company and its environment can be understood as two "part-systems" that exist in a homeostatic relationship to each other so that "the output of each is made by the other's input." The company's part-system comprises "available machinery, orders, work in progress, stocks, stores, man-power, and so on," with outputs consisting of "profit, return on capital employed, size of labour force, its earning power, the keeping of delivery promises, and so on." The environment's part-system is made up of "the level of demand, the price of money, the availability of labour, [and] the availability of materials and their price movements," and its outputs might be "the size of orders, their frequency, [and] how they are related to the reliability of previous service and to cost."[94]

The concept of self-regulating social systems built on human-computer-world metaphors that undergirds management cybernetics is exemplified in an address Beer delivered to the Pierre Teilhard de Chardin Association in October 1970.[95] In this address, Beer invited his audience to consider the global flows of commodities, labor, and capital represented in a meal eaten in a "London restaurant" with "Belgian cutlery" and "Scandinavian chairs," consisting of tinned meat from Czechoslovakia and powdered potatoes from Idaho, and served by a Cypriot waiter facilitated in his communications with the kitchen by a Japanese intercom system. He reflected on the highly dynamic character of such a world system, in which a strike in one part of the system can destabilize production, distribution, and/or consumption in some other geographically distant part. Next, in yet another example of the apparent lack of historical foresight surrounding social cybernetics, Beer turned to "those who have first hand knowledge of the fact that everything is all wrong," the "starving, the radically underprivileged, [and] workers in revolt." This account of global market dynamics and their inherent violence is remarkable for Beer's insistence that it is only "stereotypes" that lead scholars and politicians alike to consider these problems as economic or political when they are in fact cybernetic, even when no machinery in involved. In Beer's view of history, it is "unadaptive" political and economic systems, rather than the basic logic of capital itself, that produce exclusion and exploitation. The solution to these cybernetic (as opposed to socioeconomic problems), Beer insists, is to "replace *Homo Faber* with ... man the steersman—of large, complex, interactive systems." The name Beer attaches to this new political-economic subject is "*Homo Gubernator.*" In other words, Beer's solution to the emergence of a capitalist world system, with its intensification of the processes through which lives and material environments are destroyed in swathes, is to reconceptualize the human as a self-regulating node within a complex system. Finally, as if to unconsciously underscore the compatibility of this description with the emerging logic of neoliberal governmentality, Beer states that this reconceptualization must be individuated and volitional: "We must choose to become *Homo Gubernator* ourselves, while there is still time."[96]

In "Recollections of the Many Sources of Cybernetics," McCulloch recalls of Stafford Beer that he "refused several professorships, preferring to stay in business and make cybernetics pay him and industry handsomely."[97] That this account appears at odds both with Beer's self-description as a "socialist" and his well-documented work with Salvador Allende's government in Chile between 1971 and 1973 only serves to emphasize the lack of a coherent politics within the developers and adopters of cybernetic methods and the complex cunning of history that led to the deployment of these methods within the expansion and intensification of capitalist exploitation.[98] It is here worth reiterating the point that the absence of a fully articulated political-economic vision is characteristic of even the cyberneticians with designs beyond specific biological and technological research projects. Wiener, who (as has already been noted) damned the culture of "buying and selling" because it impeded free communication, was nonetheless "excited" by Beer's applications of cybernetics to management.[99] On the other side of the debates over the universality of cybernetic logic, von Neumann, although producing a body of work central to the melding of cybernetic methods with political-economic applications and being explicitly imperialist in his geopolitical outlook, never actually elaborated a theory of automata-based politics or economics (although he may well have done, completing the replacement of game theory with automata theory, had he been able to continue his work). Wiener seems to have at least sensed the growing likelihood of malign uses when he wrote in a letter to Ashby dated April 8, 1952, that cybernetics had become "infested with parasites" and that it was "so much of a field for certain individuals hopes to use it to their own advantage that I am very much put to it how to arrange my own work."[100] It must thus be seen as emblematic of the cunning of history that Stafford Beer's management theory, the objectives of which appear to have been broadly (and in the case of the Chilean project, explicitly) aligned with a socialist vision, presents a clear parallel to the types of organizational and labor practices that Deleuze, Lazzarato, Virno, Hardt and Negri, Beller, and others critique.

To support the idea that the channeling of hardware-bound cybernetic principles into industrial management is of an epistemic (rather than of a straightforwardly localized and causal) character, it is telling to examine a second system of industrial cybernetics, developed separately from Beer's in the 1950s: Jay Wright Forrester's system dynamics. Like Beer, Forrester did not participate in the Macy conferences or their offshoots. Unlike Beer, he makes no reference in his writings to cybernetics or to any of the figures discussed earlier.[101] His project appears to have been intended—more or less independently of the Macy participants and of their wider group of supporters and interlocutors—to connect the principles of feedback and self-regulation developed in servomechanical military technology to the possibilities of a probabilistic and behaviorist model of social actors and their interactions.[102]

In 1956, following the conclusion of his military work on ballistics, servomechanisms, and computing, Forrester joined the Sloan School of Management at MIT and began to develop the system dynamics approach, drawing on fundamental concepts of information-feedback systems for the study first of industry and later of more complex ecologies such as cities and, finally, the world as a whole. Although Forrester's work on industrial dynamics exhibits a terminological and representational character distinct from von Neumann's cellular automata and Beer's management cybernetics, it draws on identical assumptions about the possibility of modeling social systems by means of their conceptualization as networks of black-boxed actors that receive and process inputs and output decisions. This baseline conceptualization of the subject as a node in a network of information flows is exemplified in his book *Industrial Dynamics*, where Forrester writes that

the industrial system ... is a very complex multiple-loop and interconnected system.... Decisions are made at multiple points throughout the system. Each resulting action generates information that may be used at several but not at all decision points. This structure of cascaded and interconnected information-feedback loops, when taken together, describes the industrial system. Within a company, these decision points extend from the shipping room and the stock clerk to the board of directors. In our national economy, they extend from the aggregate decisions of consumers about the purchase of automobiles to the discount rate of the Federal Reserve Board.[103]

From this foundational system of decisional modeling, in which individual thoughts and social interactions are figured as algorithms, Forrester demonstrates the applicability of the industrial dynamics approach to the representation of advertising, order filling, manufacturing, labor flow, customer orders, and profits and dividends, with input–output diagrams presented for each.

Beer's and Forrester's configurations of workers, machinery, and capital as balanced systems of inputs and outputs produce an image of work that is conspicuous for its similarity to more recent accounts of the organization of labor under the informatic stage of capitalism. The concept of the adaptive factory in which levels of employment, wages, and activity constantly fluctuate in response to demand and resource availability must be seen as a conceptual parallel to Toyotism, the mode of production that Hardt and Negri posit as emblematic of the industrial logic of Empire under which "production planning will communicate with markets constantly and immediately" so that "[f]actories will maintain zero stock, and commodities will be produced just in time according to the present demand of the existing markets."[104] Beer's insistence that "intelligence involves as its paramount characteristic the ability to select" and Forrester's use of decision as the core organizing concept of his industrial dynamics anticipate Lazzarato's observation that post-Fordist workers are expected to function as "'active subjects' in the coordination of the various functions of production, instead of being subjected to it as simple command."[105] The same terminology used to describe the objective of homeostatic balance Beer derives from Ashby is used by Foucault in asserting the purpose of biopolitical management and

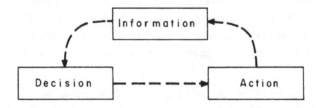

Figure 2.2
Decisions and information feedback. *Source*: Jay W. Forrester, *Industrial Dynamics* (Cambridge, MA: MIT Press, 1961), 94. Permission granted by the author.

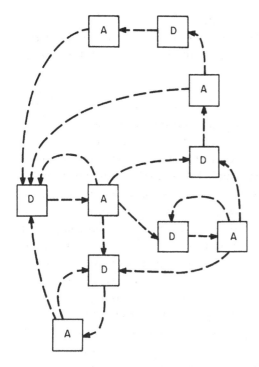

Figure 2.3
Multiloop decision-making system. *Source*: Jay W. Forrester, *Industrial Dynamics* (Cambridge, MA: MIT Press, 1961), 94. Permission granted by the author.

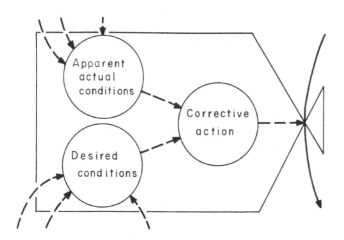

Figure 2.4
The decision process. *Source*: Jay W. Forrester, *Industrial Dynamics* (Cambridge, MA: MIT Press, 1961), 95. Permission granted by the author.

evoked in Deleuze's account of the "metastability" of the control-era business.[106] What this should make clear is that although the term *cybernetics*
itself may not appear in contemporary manuals of management theory and
industrial organization, the cunning erasure of the material specificity of
neurons and computing machines that took place around the beginning of
the 1950s is closely imbricated with the growing centrality of the principles
of that interdisciplinary field to the expansionist logic of capital.

Beyond the specific embodiments of cybernetic principles in economic
and management theory across the second half of the twentieth century, it
is ultimately the broader epistemic turn, the reformulation of the subject and
of social interaction as systems of feedback, control, and information trans-
fer, that represents the most telling endowment passed from cybernetics to
late capitalism. Even if the human–computer metaphor is only implicit in
Beer and Forrester's initial versions of cybernetic management and industrial organization, it persists as a basic condition of possibility for the idea
that labor—that is to say, the human reduced to the abstract quantity of
labor—can be conceived as and thus organized through concepts of communication and control. The central role played by (for example) the McCulloch–Pitts neuron in von Neumann's theories of computation and automata
is significant because, with human and computer collapsed into an optimized metaphor for the ideal self-regulating object of any type, the concept
of labor becomes conceptually applicable to any form of human behavior,
just as Babbage dreamed it would in the 1830s. Even if the field of cybernetics (if there ever were such a thing) dispersed into multiple distinct practices—systems theory, cognitive psychology, system dynamics, distributed
networking, supply-chain management, Toyotism, and many others—the
abstractions worked through in the 1940s and 1950s (although built on a
number of concepts that occur across a much longer period) provide these
practices with a guiding logic by framing the worker in computational terms,
as a bundle of discrete processes paralleled by the ubiquitous hardware and
software though which interactions with labor and leisure are mediated.

The historical imbrication of the concept of human labor with the idea
of the computer, which begins with Babbage and attains epistemic status
through the postwar research detailed in this chapter, points to the possibility that the "flexible" time of capitalism after cybernetics not only is
based in the breakdown of the fixed work site and the erasure of clock-
timed labor in favor of its "flexible" alternative but also can be observed
in a shift in the location of labor time from a period measured against
the linear, universal march of clocked minutes and hours to an archive of
discrete, individual *processes*. "Algorithmic time," Peter Wegner writes, "is

intentionally measured by number of instructions executed, rather than by the actual time taken by execution, in order to provide a hardware-independent measure of logical complexity."[107] Following the movement of the human–computer metaphor from cybernetics to the concepts of human behavior, cognition, affect, and social interaction, this description suggests that the "clock" that matters under control-era logics of production, whether on the floors of Wal-Mart or on Wall Street, in the cubicle or in the "Starbucks office," is the measure of workers' activity once they have been reconceptualized along computational lines. This claim is implicit—albeit without its material specificity as rooted in cybernetics—in Luc Boltanski and Ève Chiapello's book *The New Spirit of Capitalism*, where they write that "the ethic of toil, which had permeated the spirit of capitalism in various forms…tends to make way for a premium on *activity*, without any clear distinction between personal or even leisure activity and professional activity. To be doing something, to move, to change—this is what enjoys prestige, as against stability, which is often regarded as synonymous with inaction."[108] This movement from toil to activity suggests that the amount of processes completed—whether in the literal systemics of high-frequency trading or in the partly metaphorical systemics of sales targets, maintenance tasks, emails sent and answered, cold calls made, or instances of "positive attitude"—has become imaginable as a measure of time, allowing for principles of expropriation and exploitation that function at levels of granularity unthinkable in terms of hours, minutes, and seconds.

Outside of the nominal workplace (a designation that is barely stable today), the conflation of thought and action that is the legacy passed from cybernetics to capital forms the basic sociocultural principle of the communicational economy. It is not just that the keystrokes and mouse clicks (or drags, swipes, and taps) that make up web surfing constitute free labor, although this is certainly a telling indicator of the new business models at work today. It is also the principle that what one types, clicks, likes, links— that is to say, through the logic of the post-Fordist era, how one both consumes and produces at once—is socially configured as both an analog for and the totality of what one thinks, who one is, and what one will do. The necessary corollary of this vision of the social world, of course, is that any element that falls outside of this regime of positively measurable action, anything that does not constitute either symbolically representable data or an act of connection–communication, is denied existence. In the line followed from cybernetics to the society of control, one's past actions, filtered through logical formalization, become the ultimate horizon of identity, existence, and possibility.

II Digitality as Cultural Logic

Part I of this book outlined a genealogy of so-called digital culture by examining the ways in which the computer functions both as a tool and as a universal metaphor closely attuned to the logics of equivalence, expansion, and subsumption inherent to capitalism. The concept of the control episteme—which links the principles of post-Fordist production, communicative and cognitive capitalism, and neoliberal economism through nested concepts of biology, psychology, markets, and society as self-regulating systems premised on data processing—was established as a central principle upon which ideals of informatic, always-on, and globally distributed modes of organization and valorization are premised. This principle, though grounded in technological objects and the practices they facilitate, is both socially and culturally shaped and has both social and cultural implications that can be articulated as follows:

1. The epistemic reconceptualization of the human as a computing machine grounds assumptions that cognition, attention, communication, and creativity are fully intelligible and representable as digital systems (and thus as forms of value-producing labor) and that flexibility and constant activity are behavioral norms.
2. The spatiality and temporality of labor, formerly bounded by the fixed site of the workplace and a clear distinction between labor time and non-labor time, or production and reproduction, expand under the images of the human as computing machine and the social as digital communication system. In the overdeveloped world, this expansion is manifested in phenomena such as the twenty-four/seven time–space of labor at the higher end of the socioeconomic spectrum and the zero-hours contract farther down the spectrum. In the underdeveloped world, the expansion is manifested in brutal working conditions and vast spaces of material dispossession that both result from and exist in tension with the optimized image

of the human and the world as arrangements of immaterial information processors. Neferti X. M. Tadiar's account of the distinction between the creative worker and the "ideal migrant domestic worker," who "from the side of her employment ... images the archetypal robot, capable of offering emotional, as well as menial, help to humans, without expectation of human feeling in return," is emblematic in this regard.[1]

3. The computational vision of the social actor and the systemic vision of the world, though built on idealized notions of computing machines, are necessarily finite and synchronic even as they ground claims to totality and universality (see Oliver, Pierce, and Shannon on PCM and Kittler on computing in the introduction to this book as well as von Neumann and Ashby on the black box in chapter 3).

4. These conceptualizations thus introduce a series of breaks into the field of the living, where immobility, fixity, and disconnection from channels of communication appear aberrant or pathological and thus lead to expulsion from circuits of representation and inclusion. The conditions that lead to nonrepresentation and expulsion function across a multitude of nested layers: they apply both to individuals and groups and to particular aspects of thought, identity, behavior, desire, and so on at the level of the dividual.

In delineating the emergence and implications of the control episteme, part I posited a particular mode of critical analysis. This mode of analysis requires paying close attention to specific historical events and material technologies within a wider, historicist approach to the epistemic and cultural valences that emerge in concert with certain concepts and technologies. Building on these theoretical and historical foundations, the chapters in part II evaluate the penetration and power of the control paradigm from the mid–twentieth century on by examining some ways in which they are imbricated with, constituted by, and represented across a range of cultural objects and practices.

In "The Cybernetic Hypothesis," Tiqqun make the following claims about the imbrications of cybernetics, economic theory, and governance:

The preponderant role that cybernetics was to play in the economy after 1945 can be understood in light of Marx's intuitive observation that "in political economy the law is determined by its contrary, that is, the absence of laws. *The true law of political economy is chance.*" In order to prove that capitalism was not a factor in entropy and social chaos, the economic discourse gave primacy to a cybernetic redefinition of psychology starting in the 1940s. It based itself on the "game theory" model, developed by von Neumann and Oskar Morgenstern in 1944. The first sociocyberneticians showed that *homo economicus* could only exist on the condition that there would be a *total transparency* of his preferences, regarding himself and others.[2]

In making these claims, Tiqqun delineate the contours of an inquiry into the relationship between control and cultural form. First, they observe that the warding off of chance that drives the control episteme (and that can be observed through both the systematization of work processes and the deployment of statistical modeling and forecasting) can be produced only through (1) a conceptualization of the individual as a black box, a bundle of discrete, measurable traits and behaviors whose internal workings are deemed surplus to the requirements of any meaningful analysis even as they are understood as unknowable due to their opacity, and (2) a conceptualization of society as a network of such black boxes, each of which acts (or is invisibly motivated to act) in ways that can be captured and defined within predetermined grids or databases. This is nothing less than a sociotechnical account of neoliberal orthodoxy: there is no such thing as society, only individuals and symbolically formulated interactions subject to probability-forecasting equations. Second, Tiqqun foreground the necessity of *"total transparency"* and in so doing suggest a valence of the cybernetic hypothesis of capital that must be traced through cultural forms and practices as well as sociotechnical phenomena (if these categories can even be separated). As writers from Guy Debord to Jonathan Beller have shown, the construction of systems of total transparency based on political-economic fantasies of control tend to be organized around text, sounds, and images even if their functional bases resolve to pure data.

Following Tiqqun's provocation, the cultural formations that shape and are shaped by the logic of control-era capital can be traced in two directions: on the one hand, there is the production of aesthetic regimes that directly stimulate interaction and thus facilitate the processes of data collection that are necessary to construct models of the social interaction as a bundle of totally transparent processes (Facebook's social graph is exemplary of this tendency); on the other hand, there is the production of aesthetic objects that attest to the ways in which actors internalize the processes of capture and control, which then shape their cognitive-linguistic capacities and thus their cultural production. This second category might be understood as the political unconscious of the control society. To critically engage with these intertwined aesthetic regimes, it is necessary to simultaneously engage digital technologies and the cluster of (often aesthetic) metaphors that surround them. The distinct parts of this complex are brought together in a passage by Geert Lovink and Florian Schneider, although the outcome of their inquiry is quite different to the one pursued here. In an essay titled "Notes on the State of Networking," Lovink and Schneider reflect on the ways in which the critical power of Debord's *Society of the Spectacle* has been undercut by the networking paradigm of informatic capitalism, which "escapes the centrality of the icon to visual culture and its critics and instead focuses on more abstract, invisible, subtle processes and feedback loops." "There is nothing spectacular about networking," Lovink and Schneider insist; thus, "most of the leading theorists are not aware of the current power transformations" because they "still sit in front of the television and watch the news or a rental VHS."[3] The central claims here—about the necessity of engaging with the "invisible, subtle processes and feedback loops" of the networking paradigm and the fundamentally nonspectacular character of networking—are compelling, but Lovink and Schneider discard visual and narrative form too easily. Such an approach fails to engage with the ways in which the motivation of volitional human action upon which networked accumulation with its feedback loops and "invisible" processes is built requires constant stimulation. This is why the most recent media technologies, from web pages to video games, are stimulants, whereas their noninteractive predecessors were opiates. To begin addressing the social and cultural logics that control motivates, then, it is necessary to understand not only the informatic conceptualizations of actors and groups, but also the specific cultural modes that both produce and are produced by these conceptualizations.

Addressing the cultural resonances of control does not equate to retreating from its material socioeconomic and environmental implications, from exploitative labor and the valorization of identity and sociality to exclusion, poverty, and resource depletion. Because of the fractal character of cybernetic logic that is evidenced in its conceptual scalability from neuron to social actor to factory to world, it follows that the intensification of capitalism's cybernetic properties will be principally traceable across (and at the crossing points of) multiple technical, social, and cultural layers. As theorized in part I, this conceptual scalability requires procedures of idealization and exclusion and thus rests on conceptual formulations that erase the materiality of neurons, computers, bodies, and material resources alike. Therefore, the severest socioeconomic implications of control must be seen to be both (1) founded in the logical and technical principles of computing machines and (2) constitutive of a political unconscious that is expressed in those cultural forms that are produced under and alongside the expanding conditions of capital that draw a conceptual and technological endowment from these machines. Chapter 3 elaborates these principles through an analysis of the critical challenges posed to representation in the age of control and in so doing establishes the methodological concerns that the remaining chapters will attempt to engage and move beyond.

Chapter 4 traces the emergence of control's cultural valences from the 1940s to the 1980s through an examination of the technical, conceptual, and cultural relationships between text and computation. It works through the emergence of control as a cultural as well as a technical and socioeconomic logic, while also elaborating an interpretive methodology for the location of the technical and socioeconomic at the level of the literary. Finally, chapter 5 examines the narrative logic that digital conceptualizations of the subject produce, moving toward a mode of interpretation attuned to tracing the socioeconomic and cultural valences of control in contemporary cultural artifacts such as films and video games.

3   On Representation

Exclusion and Unmeasurables

In "Life-Times in Fate-Playing," Neferti Tadiar pursues a rigorous analysis of the totalizing concept of life that inheres in many critical analyses of real subsumption and post-Fordism. After working through (and finding much of value in) the critique of real subsumption that is foundational to concepts of immaterial labor and cognitive capitalism presented by Antonio Negri, Paulo Virno, and Christian Marazzi, Tadiar queries the potential for these theorizations to undergird a "broader and more concrete antiracist, postcolonial feminist consideration of the potential of 'living labor.'"[1] The universality of "life" as a concept in these critiques is problematic for Tadiar because it is "indifferent to the qualitative gradations and divisions of 'life' that obtain as crucial aspects of contemporary processes of value extraction."[2] Tadiar's analysis suggests that the metaphorical concept of life as a homogeneous source of value that constitutes the worldview of post-Fordism (if not of capitalism in general) is inadvertently adopted and thus naturalized in critiques of the latter. This adoption can be read as an example of the succession of epistemic victories that characterizes the emergence of control. The totalizing concept of life as labor fails to acknowledge the facts that under capitalism "not all life is valorized or valorizable," that "the extraction of value from 'life' takes place through more than one modality," and that "life' in the complex dynamics of capitalist processes assumes radically contradictory forms."[3] Tadiar's intervention marks an essential element of the analysis of control pursued here. First and foremost, it locates the same absence of a temporal dimension in critical accounts of post-Fordism that Maya Deren finds in game theory. Within such totalizing concepts as post-Fordism or the real subsumption of life under capital, there is no accounting for the exhaustion of life over time, which Tadiar identifies as the "used-up" life of the "unskilled" Mexican or Chinese woman worker in contrast to that of the "life entrepreneur" or "creative worker."[4] If the expansion and universalization of the notion of valorizable time composes a core component of the cybernetic logic grounding such historical shifts as that from Fordism to post-Fordism, its reproduction in the same critical worldview that seeks to reveal and overturn it can only be read as a slippage that illustrates the epistemic power of control. Identifying the tension between specific material practices and the vapory notion of the computer as universal metaphor represents one way of addressing the contingency of this slippage, but the depth and breadth of control's penetration means that to properly apprehend it requires many more inquiries into social and cultural practices.

Alongside the standardization of life that both post-Fordism and its critiques rest upon, Tadiar evokes the "redundancy" and "superfluity" of vast swathes of what can be understood as lives and aspects of life. In addition to the "wasting process" effected on the lives of ostensibly low-skilled workers in the underdeveloped world, Tadiar observes a logic of exclusion that, although inherent to capitalism in general, is intensified by the flexible (or precarious) working conditions of post-Fordist production: "The expanded reproduction of capital depends on the movement of labor in and out of lines to offset falling profitability, through the 'setting free' of workers compelled to find employment in expanding markets. Although this pool of newly released, discounted (at fire-sale prices) labor is reabsorbed by capital as an industrial reserve for the regulation of the labor market, it tends to outgrow this function and to become 'a consolidated surplus population, absolutely redundant to the needs of capital.'"[5] As the imperative of constant valorizable activity across multiple spaces and time scales emerges as a basic condition of productivity, it determines the basic conditions of existence in general in new ways. If capital can only apprehend life as labor, as Marx writes in the *Grundrisse*, the expansion of labor through real subsumption will necessarily expand certain conditions of productivity and thus existence from the point of view of capital while deepening the conditions and implications of nonproductivity and nonexistence. As Saskia Sassen puts it in relation to a specific deployment of this logic of ontological redundancy, "The unemployed who lose everything—jobs, homes, medical insurance—easily fall off the edge of what is defined as 'the economy,'" while leaving overall measures of growth untouched. Systemic exclusion thus "makes 'the economy' presentable … allowing it to show a slight growth in its measure of GDP per capita" in a procedure "akin to a kind of ethnic cleansing in which elements considered troublesome are dealt with by simply eliminating them."[6]

If the expanded reproduction of capital is—as this book argues it is—at least partly grounded in an epistemic shift that reformulates concepts of the human and sociality around computational metaphors and that emerges in cunning ways from the expanded cultural field surrounding cybernetics research, then it follows that the production of redundancy, in this case surplus populations, is a component of the same technical and logical base. Like McCulloch and Pitts's ideal and dematerialized neuron or Turing's universal machine, the life of the laborer that is conceptualized through the industrial models formulated by Beer, Forrester, or Toyota or the social graphs of Google or Facebook is stripped of much of its material, social, and psychological specificity, all of which are conceived of as outside of even the expanded field of life-as-production and thus rendered invisible, unworthy of being attributed systemic representation and, by extension, existence. In this world picture, there is no thought, identity, affect, or sociality outside of processes of communication, with the latter understood not as anything to do with community but rather as a sequence of discrete, countable steps. The conceptualization of the material and social world through the metaphor of the digital computer as idealized information processor and communication system is thus premised on processes of division and exclusion that scale from what Friedrich Kittler calls the "impossibility of digitizing the body of real numbers formerly known as nature"[7] to what Tadiar calls the "consolidated surplus population, absolutely redundant to the needs of capital."

If one accepts that the control-era mutation of capitalism (of which post-Fordist production is one valence) draws a significant endowment from cybernetics, then one should not be surprised to find this principle of necessary exclusion at the heart of that discipline's epistemic turn. The concept of the black box exemplifies the basis of a methodology in which all interest in internal function is discarded in favor of measurable behaviors understood as inputs and outputs.

In winter 1943—before the field of cybernetics was named, before the Macy conferences, but after the cerebral inhibition meeting (see chapter 2), Bigelow and Wiener's antiaircraft prediction work, and McCulloch and Pitts's publication of "A Logical Calculus"—a group of mathematicians, engineers, and psychologists, many of whom worked on the earliest experiments and publications in what would later be called cybernetics, were gathered together at Princeton and presented with a thought experiment that would become central to the universalization of cybernetic logic. The participants in the Princeton meeting—including Wiener, von Neumann, McCulloch, and the neuroscientist Rafael Lorente de Nó—were asked to consider a pair of hypothetical black boxes.[8] They were told that these boxes had been liberated from the German army and that the first box had exploded upon attempts to open it. McCulloch recalls that "both [boxes] had inputs and outputs, so labeled," and that the exercise those present were given was "phrased unforgettably: 'This is the enemy's machine. You always have to find out what it does and how it does it. What shall we do?'"[9] Although the outcome of this experiment appears to have been inconclusive, the power of the black box as concept can be seen to have a lasting effect on the later development and epistemic diffusion of cybernetic logic.

The epistemic function of the black box concept begins to become apparent in von Neumann's "General and Logical Theory of Automata." In this talk, von Neumann, while discussing prefatory methodological concerns, explicitly states that the treatment of individual elements as "black boxes" represents a necessary precondition for a universal theory of biological and technological systems. This procedure of "black boxing," von Neumann clarifies, involves treating the key elements within a system as "automatisms[,] the inner structure of which need not be disclosed, but which are assumed to react to certain unambiguously defined stimuli, by certain unambiguously defined responses."[10] Von Neumann returns to the black box metaphor on four further occasions in "General and Logical Theory of Automata," to discuss computing machines first, then neurons, and finally a general automaton. Within the historical inquiry developed here, it is the last that is most telling in terms of the logical substrate of post-Fordism or the industrial register of control:

An automaton is a "black box," which will not be described in detail but is expected to have the following attributes. It possesses a finite number of states, which need be prima facie characterized only by stating their number, say n, and by enumerating them accordingly: 1, 2, . . . n. The essential operating characteristic of the automaton consists of describing how it is caused to change its state, that is, to go over from a state i into a state j. This change requires some interaction with the outside world, which will be standardized in the following manner. As far as the machine is concerned, let the whole outside world consist of a long paper tape.[11]

Although von Neumann is describing Turing's account of a computing automaton, it is not difficult to locate the information-processing concept of the social actor in his account. The black box, within the history of cybernetics and thus within the epistemic grounding of control, emblematizes the methodological principle that elements inherent and/or internal to a given object—be it understood as a machine, a biological organism, or a social being—can be excluded in favor of statistical methods based on inputs and outputs (or stimuli and behaviors) without sacrificing any explanatory or diagnostic facility. The conceptualization of action and interaction (between neurons, logic gates, or social actors) as communication, for example, relies on this principle.

Within the historical diffusion of the black-box principle, it is suggestive that the metaphor is especially central to W. Ross Ashby's book *An Introduction to Cybernetics* because that is perhaps the first text to fully exhibit the elision of the distinction between applications of cybernetics to biology, technology, and economics. Ashby claims that his book, in which he devotes an entire chapter to a discussion of the black-box concept, will demonstrate "essential methods" that can be "used uniformly" to "attack the ills—psychological, social, economic—which at present are defeating us by their intrinsic complexity."[12] The principle that these methods require the exclusion of any element that poses problems to discrete representability is already visible in Ashby's previous book, *Design for a Brain*, published in 1954. In that text, Ashby, whose homeostat (an adaptive, self-stabilizing machine model of the brain) had been demonstrated at the ninth Macy conference in March 1952, sets out a series of procedures for the modeling of human and animal behavior based on the principle that "a machine or animal behave[s] in a certain way at a certain moment because its physical and chemical nature at that moment allow[s] it no other action" and that therefore any systemic model of human affairs must deal "not with qualities but with behavior ... [and] not with what an organism feels or thinks, but with what it does."[13] Following this claim, Ashby underscores the principle of prediction that is fundamental to the economic adoption of cybernetics by writing that "it will be further assumed ... that the functioning units of the nervous system, and of the environment, are all strictly determinate: that if on two occasions they are brought to the same state the same behavior will follow."[14]

The idealized prospect of socioeconomic control elaborated by Ashby in *Design for a Brain* and *An Introduction to Cybernetics* is emblematic of the universalized version of cybernetics that this book argues provides an epistemic grounding to late capitalism. Equally, it is clear that the principle of exclusion, of which the black box is exemplary, is central to this construction of cybernetics and thus to the formation of control. For example, although behavior provides Ashby with a ground for modeling the brain, he explicitly excludes consciousness, writing: "If consciousness is the most fundamental fact of all, why is it not used in this book? The answer, in my opinion, is that science deals, and can deal with, only what one man can *demonstrate* to another. Vivid though consciousness may be to its possessor,

there is as yet no method known by which he can demonstrate his experience to another."[15] The inclusion of this particular example is not intended
to ground a romantic gesture, the holding up of unique, sovereign, human
consciousness as a privileged realm against the base impositions of positiv-
ism. Rather, it is included to foreground the centrality of exclusion to cybernetic modeling. From the cyberneticians' point of view, the impossibility
of including certain factors in a model does not invalidate the methods or
the very possibility of useful socioeconomic modeling but rather marks the
limits that must be placed on the world for such modeling to be applied.
Anything that cannot be included is denied functional existence, whether
it is consciousness or the body of the maquiladora worker no longer able
to execute productive actions. The black-box concept exemplifies the historical construction of the computer not only as a tool but also as a source
of vague metaphors—or what Hans-Jörg Rheinberger calls an "epistemic
thing," a material entity that presents itself in a "characteristic, irreduc-
ible vagueness"[16]—that forms a logical precursor to the particular, systemic
processes of valorization and exclusion that characterize control-era capital.
The meeting points between these epistemic things and the concrete social
practices they produce are central to the most pressing questions pertaining
to representation in the age of control.

Representing Control

How might one locate subjective representations of the cybernetically modeled world, that flat ontology of black boxes with its fractal processes of
exclusion and erasure? When one is engaged with questions of global representation in the era of ubiquitous digitality, the technical and historical
imbrications of cybernetic systems with the distributed network form would
appear to be central. It is clear that the network diagram is a dominant mode
of modeling and thus of understanding social practices today. From the distributed communications architecture proposed by Paul Baran to the RAND
Corporation in 1964 to the emergence of the World Wide Web in the early
1990s and the rise of social networking sites such as Facebook and Twitter,
electronic communications have clearly played a major role in fostering the
current ubiquity of the network as both infrastructure and method.[17] With
the extension of discretized symbolic logic beyond technical computing
machines and into models of social management, the network form as a
mode of representation is not limited to computer-mediated interactions.
Today military strategy deploys the insights of operations research and system dynamics to frame a given conflict as a network comprising soldiers,

ordnance, and contested physical space as well as civilian infrastructures, belief systems, communications systems, and global news coverage, among many other facets.[18] In business strategy, books and consultancy reports abound heralding the benefits of understanding flows of money, services, resources, information, and influence as networks.[19] Geoff Mulgan, a director of policy under the New Labour government, advocates network-based concepts of politics in his books *Communication and Control* (1991) and *Connexity* (1997), and the report *Adding It Up* produced by the British government's Performance and Innovation Unit in 2001 opens with a call from the then prime minister Tony Blair to "ensure that analysis and modeling is given due weight in policy advice ... and that analysis, like policy itself, is properly joined up."[20] Theoretical approaches to the social—including Bruno Latour's Actor-Network Theory (ANT), Manual Castells's book *The Rise of the Network Society*, Albert-László Barabási's book *Linked: The New Science of Networks*, and Yochai Benkler's book *The Wealth of Networks* as well as digital humanities applications such as Franco Moretti's diagrams of social interaction in *Hamlet*—all attest to the role the network form has played over the past two decades as an apparently universal model for the analysis of complex social phenomena.[21]

Following the misgivings about cybernetic logic as an appropriate mode for representing sociality worked through in part I, however, it is necessary to look beyond the apparently objective role of the network form in the various accounts described here. The commonsense notion that native digital representational modes—that is, the types of diagram found in work by McCulloch and Pitts, Beer, and Forrester; the more recent network diagrams touched on earlier; and computational forms such as algorithms, software applications, graphical user interfaces (GUIs), and video games—are the only way to make sense of the cultural logic of control must be problematized if the broadest implications of digitality are to be confronted. In each of these forms of representation, only black boxes (standing in for neurons, computers, workers, or what have you) and their interconnections (or inputs and outputs) can be included. Anything "inside" the box or outside of the categories of input or output is left to fall out of representation altogether, a fate that directly scales up to the dispossession of those forms of life that Tadiar identifies as unvalorizable under the current conditions of the global, networked, and flexible stage of capitalism.

Although computers, software, and networks can reveal a great deal about the internal logic of a control society, their historical entanglement and isomorphism with this logic means that they are, in isolation, precluded from revealing many of the material impositions and exclusions of such a society. To restate one of the fundamental assertions of this book, it is the tension between digital-symbolic representations of the social and the continuous, rich and multiple experience of actual social existence—as well as the very real forms of subjectification and exploitation this tension occludes—that must form the central concern of any investigation into the cultural logic of the digital. Therefore, to assert the cultural primacy of digital representation is to ignore the prospect that when it comes to the specific practices of critical analysis that control necessitates, mutations in forms inherited from prior eras may tell one as much about the cultural-political terms of this era as an examination of its native objects does.

Before one begins to address the ways in which representations of the control era that take into account both systematicity and exclusion might be located, it is necessary to consider some specific methodological concerns, and Fredric Jameson's concept of cognitive mapping serves as a useful focal point for such an evaluation. Developed in the 1980s and early 1990s, the concept of cognitive mapping describes the process through which a subject might make sense of the relationship between their immediate local position and the global late-capitalist networks, which, because of their scale and complexity, cannot be conceived of as a whole, let alone adequately represented in their totality by available modes of aesthetic expression.[22] Built on a synthesis of the Althusserian concept of ideology and aspects of the geographer Kevin Lynch's book *The Image of the City*, the cognitive map, for Jameson, is a method for making a properly unrepresentable totality at least partially legible. It is a method that enables one to "think a system so vast that it cannot be encompassed by the natural and historically developed categories of perception with which human beings normally orient themselves." The fundamental constituents of such a map, Jameson asserts, are "space" and "demography," even if these categories are to be "used like a ladder and kicked away" once they have done their critical "work."[23]

In Jameson's account, cognitive mapping entails the following procedure: (1) a more or less allegorical representation of a system and (2) a representation of the way the local elements of this system relate to the global totality are produced, so that (3) a reader, viewer, or user can grasp the otherwise unimaginable relation between the two. This multivalent aspect of the cognitive map, whose fundamental promise is to provide a way to think through the connection between experience and global social systems and which appears in Jameson's own work most commonly in relation to cinema, is also the source of the difficulties posed to the procedure by digital media. Cognitive mapping describes a mode of representation that seeks to overcome a problem of scale in which the relationship between the parts and the whole of a given system poses the fundamental problem to be resolved. As such, the procedure of cognitive mapping that Jameson establishes as essential to the global system of financial capital bears an uncanny similarity to the network diagram that rose to ubiquity from the early 1990s on. That this diagram is emblematic of the control vision of the world as a series of nodes connected only by lines of communicative action foregrounds the historical problems surrounding its use as a method for grasping the wider socioeconomic implications of this worldview.

The problem posed by this commensurability between the "classical" procedure of cognitive mapping and the network mode of representation is apparent in recent work by Wendy Hui Kyong Chun and Alexander Galloway. In Chun's book *Programmed Visions*, the cognitive map is evoked in relation to the network diagrams of economic and power relations produced by artist-activists such as theyrule.net. Chun writes that the distributed network diagram, whether accessed or composed through user-centered computer interfaces, facilitates the "difficult and necessary" procedure of mapping "invisible networks" of capital because the creation of such diagrams produces "'informed' individuals who can overcome the chaos of global capitalism by mapping their relation to the totality of the global capitalist system."[24] Chun's analysis, while containing sharp insights about the way software-centered computing mirrors both ideology and ideology critique, also foregrounds the possibility that the diagrammatic power of the network diagram functions to foreclose the subjective dimension of the cognitive mapping procedure, and in so doing robs this procedure of its critical efficacy. In other words, although Chun's analysis indicates a historical progression whereby the formerly invisible network of global capital whose representation constitutes one pole of the cognitive mapping procedure becomes at least temporarily visible, it does not move one any closer to grasping the immanent or subjective experience of these now somewhat visualizable systems that compose the dialectically linked second pole of the mapping procedure. In and of itself, the procedure through which a complex system is represented as a network of relations between nodes—be they human actors, commodities, resources, ordnance, or something else—describes little more than the logic that underpins control societies, wherein only positively measurable behaviors can be relied on to model social systems as well as technical ones. This form of mapping must thus be seen as grounded in the historical logic of control. Although the representability that this form extends to the circulation of capital might appear to obviate the need for the more allegorical modes of cognitive mapping such as those Jameson locates in films like *Dog Day Afternoon* (Sidney Lumet, 1975), the network or system diagram fails to account for the necessary processes of exclusion, optimization, and filtering that make such representations possible in the first place. In short, the network diagram is not an objective depiction of relations but rather a depiction of the logic that produces them by excluding everything that falls outside certain categories or below certain thresholds.[25]

As argued throughout this book, the discrepancy between what can be modeled and systematized and the sum of what might exist is central to critiques of the cybernetic sciences as a blueprint for control-era social logic. From Warren McCulloch's 1974 claim that although cybernetic approaches provide a compelling framework for electrical engineering and neuroscience, their application to society is "mere suspiration" to Tiqqun's extended critique of the cybernetic roots of neoliberalism, which they observe is premised on "hollow[ing] out" the subject, disqualifying "individual inwardness/internal dialogue, and with it all nineteenth-century psychology, including psychoanalysis" as a "myth,"[26] the inadequacy of the distributed network form for mapping the social arises from the filtering out or erasure of unmeasurable components of being that is at the heart of its methodology. Galloway explicitly formulates this as a problem of representation in his essay "Are Some Things Unrepresentable?," writing that network diagrams "offer us no orientation whatsoever within the social totality" of control societies and that this inadequacy rests in part on the totalizing representational mode that such diagrams inhabit: network diagrams and other information visualizations not only fail to allow us to orientate ourselves but also "exacerbate the problem by veiling it behind candy-colored lines and nodes."[27]

Galloway's analysis of the troubled relationship between networks and cognitive mapping is compelling for two reasons: it is rooted in the present ubiquity and homogeneity of distributed network diagrams, which now simply reformulate (rather than solve) the problem of representing the scale and complexity of social systems; and it points to the necessary future project of finding modes of mapping and representation that can circumvent the totalizing indifference of informatic systems. These interrelated components come together in the concluding claim that

algorithms and other logical structures are uniquely, and perhaps not surprisingly, monolithic in their historical development. There is one game in town: a positivistic dominant of reductive, systemic efficiency and expediency. Offering a counteraesthetic in the face of such systematicity is the first step toward building a poetics for it, a language of representability adequate to it. Thus while unearthing alternatives might seem challenging, once the first few steps are taken one witnesses a wide-open plane, a vast anti-history of informatics waiting to be written, a vast world of representation waiting to be inscribed. To create a poetics for such algorithmic systems is the first step, necessary but not sufficient, in the quest to *represent* them.[28]

The search for such a poetics does not simply entail switching from "passive" to "interactive" representations. The principal concerns raised in Galloway's critique of information aesthetics are also applicable to interactive digital media such as video games. In a chapter of his book *Gaming* that addresses allegories of control in the strategy game *Civilization*, for example, Galloway argues that the fundamental digital logic of video games extends to their modes of interaction even if the latter ostensibly involve continuous bodily movements on the part of the player. Algorithmic logic technically precludes video games from representing any outside to informatic models, as anything that appears in them must by definition be representable in digital form. "'History' in *Civilization*," Galloway suggests, "is precisely the opposite of history [that is, 'the slow negotiated struggle of individuals together with others in their material reality'] ... because the diachronic details of lived life are replaced by the synchronic homogeneity of code pure and simple."[29] This rendering of lived life by synchronic code is a result not of the imperialist politics *Civilization* appears to celebrate but rather of the totalizing logic that the game's very nature as a digital system makes inescapable—as Galloway puts it, even a "politically progressive '*People's Civilization*' game, à la Howard Zinn[,] would beg the same critique."[30]

Although both Chun's and Galloway's accounts foreground the problems posed to the process of cognitive mapping by computer networks and the social situations they both enable and model, neither moves toward a method of analysis commensurate with the material implications that accompany the representational modes of control. In order to properly grasp control's socioeconomic and cultural costs, it is necessary to account for the computer's metaphorical as well as literal roles in the development of new processes of exploitation and subsumption. It will be through the development of such an analysis that the necessity of noncomputational forms (in the forms they take once computers, cybernetic systems, and the network diagram have become ubiquitous) will be made clear. The mode of cultural analysis that control necessitates is thus one that takes the monolithic historical dimension of algorithmic or networked logic (as sketched out in the discussion of the cybernetic sciences set out earlier) as inseparable from the formations of sense and subjectivity that produce cultural forms. The representational modes that this analysis attends to might not resemble Jameson's cognitive mapping, in which *space* and *demography* are the operative terms, so much as an affective mapping of the tension between demography (or any number of other discrete systems of modeling and management) and the limitations of the systems that are deployed to model it.[31]

Aesthetic Demography

The contours of an approach to interpretation in which complex procedures of subjectification and desubjectification are seen to constitute the social actor engaged with networks of commodity flows (of which both interactive and noninteractive commercial media are emblematic) are sketched out in a passage in Giorgio Agamben's essay "What Is an Apparatus?" in which he writes

What defines the apparatuses that we have to deal with in the current phase of capitalism is that they no longer act as much through the production of a subject, as through the processes of what can be called desubjectification. A desubjectifying moment is certainly implicit in every process of subjectification.... But what we are now witnessing is that processes of subjectification and processes of desubjectification seem to become *reciprocally indifferent,* and so they do not give rise to the recomposition of a new subject, except in larval or, as it were, spectral form.... He who lets himself be captured by the "cellular telephone" apparatus—whatever the intensity of the desire that has driven him—cannot acquire a new subjectivity, but only a number through which he can, eventually, be controlled. The spectator who spends his evenings in front of the television set only gets, in exchange for his desubjectification, the frustrated mask of the couch potato, or his inclusion in the calculation of viewership ratings.[32]

Although Agamben's focus here is on mobile telephones and television, it is not difficult to identify a broader argument about literal and metaphorical digitization, capital, and exclusion that is of central concern in this chapter. Two aspects of Agamben's argument are especially pertinent for the present discussion of representation and control: the first is the replacement of the concept of the subject by data models such as those he somewhat vaguely refers to as "viewership ratings"; and the second is the association of indifference to subject formation that this process exhibits. These interlinked concepts of data and desubjectification can be seen to structure the relationship between control and aesthetics in a way that allows the latter to be read as indicative of the social and economic penetration of the former. The practices of demographic modeling that determine the aesthetic mode of contemporary commercial media production point to a particularly explicit layer of the processes through which the cybernetic logic of contemporary capital becomes visible and thus traceable in contemporary cultural production.

Of course, forms of demographic targeting have always played a major role in the production and marketing of commodities. What distinguishes the control mode from its predecessors is the detail and intensity of contemporary modeling systems, which no longer focus on the "four quadrants" of classical demographic targeting (male, female, older than twenty-five, and younger than twenty-five) but view the social field as a dynamic but nonetheless controllable system of interacting predicates including broad interests, social and political outlooks, "values," and "lifestyles" as well as race, gender, income, and geography. The specific tendency toward conceiving of the consumer as a node in a communications network, with the associated deployment of multiple media as inputs rather than as ends, is perhaps nowhere better emphasized than in a call to arms by Jeff De Joseph of the advertising agency Doremus, published in the September 1999 issue of *Adweek* magazine. Reflecting on the advertising field's increasing informatic complexity just as Doremus is about to launch that staple of the control episteme, a "new mathematical model that will bring order out of the techno-chaos," De Joseph sets out a number of procedures that the advertising community must follow in order to grasp a discrete, logical order below the apparent complexity of social reality:

• Understand that to consumers, virtually all categories are commoditized. Products become interchangeable. The defining component is how the vision and values are expressed.

• Learn from dot coms that high-impact, holistic marketing makes non-entities into macro-brands. The mission of marketing is to surround the market with the brand's proposition, values and differences. And the best way to do this is through the style and sophistication of the communications.

• Take advantage of leverage. The audience needs to be refreshed by new communications. Nothing stands still, especially the dialogue with the street. Even if people don't know the product, they can recite the slogan. Demonstrate vitality and viability using communication as a surrogate.

• Take advantage of the media's need to feed itself [sic]. Create content. Spread the word through print, TV and the Web.[33]

Several things are immediately notable about De Joseph's commandments. First, none of them is predicated on the specific function of any particular technology. They all describe modes of seeing and understanding rather than technological processes, and they function prior to and alongside the type of mathematical modeling and prediction system that Doremus promises to offer. Second, they begin from the understanding that consumers are not blind to but rather directly recognize the interchangeability of commodities and that product differentiation is about the forms of knowledge that surround a given commodity rather than about material variation. Vision, values, and the style and sophistication of communications differentiate brands as distinct from physical products (and here it is easy to see a distinction along the lines of the one Deleuze posits between the factory's material processes and outputs, on the one hand, and the "soul" or "gas" of the business, on the other). Third, De Joseph's procedures conflate communication and life, emphasizing the refreshing power, vitality, and viability of the former and positing "the media" as a living system that requires communicative input to sustain its negative feedback loops. Finally, this call is remarkable for its title, "Beyond Knowing," which articulates the cybernetic elevation of possibility and statistical forecasting over the deterministic models and fixed categories associated with the supposedly outdated concept of "what can be known."

This communicative model of advertising, which purports to be effective because it does away with the rigidity of fixed relations in favor of flexible, dynamic models, is nonetheless predicated on ideals of definition and categorization that sit in a genealogy encompassing: Babbage's workplace organization, which assigns more or less mundane and dangerous tasks based on the worker's assumed ability; Hollerith's computation-by-essentialization of the US populace in 1890; and the various cybernetic models of social interaction developed from the 1940s onward that are detailed in chapter 2. In *The Cultural Logic of Computation*, David Golumbia places this type of dynamic definition and targeting alongside management-system regimes such as customer relationship management and enterprise resource planning (each of which owes a clear debt to the cybernetic disciplines of operations research and system dynamics) and under the banner term *cultural striation*. Focusing on Claritas Corporation's PRIZM system, Golumbia demonstrates the basic principles of definition that facilitate even the most apparently dynamic, flexible approaches to audience targeting. Based on Jonathan Robbin's concept of "geodemography," PRIZM "divides the U.S. consumer into 14 different groups and 66 different segments." Groups include "Urban Uptown" and "Urban Cores," and segments within these groups include "Young Digerati" ("tech-savvy singles") and "Big City Blues" ("downscale Asian and African-American households occupying older inner-city apartments" within the "Urban Core" group.[34] Systems such as PRIZM purport to offer a picture of social reality that is able to move beyond the limits of simple demographic categories by including preferences and conditions of knowledge such as tech-savviness, but fall short of the full-spectrum control that De Joseph's "Beyond Knowing" presents as the ideal of the information age demographer because they remain necessarily tied to a finite number of cores and segments. This tension between finite definition and fully flexible capture points to the way in which control exists as a set of ad hoc practices grounded in but always destined to fall short of the ideal of the world as computational system.

PRIZM is far from the only system that promises rich social visualization based on the interaction of fixed and amorphous predicates (indeed, Doremus's "new mathematical model" detailed above is another, even if it is grounded in the totalizing ideal De Joseph presents), and it is evoked here to exemplify a particular tendency rather than to single it out for special attention. The same logic is applied to a wider range of cultural markers in Urban Outfitters' conceptualization of the "model" customers of each of their lines:

[Urban Outfitters]: The upscale homeless person, who has a slight degree of angst and is probably in the life stage of 18 to 26 ... the incoming freshman, the struggling art school kid, the girl who dresses differently than her friends. The kid who has a band, anybody who has a band is our customer ... And the girl who has not realized that quirky is sexy and that being a hipster is not simply a marketing tool, it's someone that just does something differently than others.

[Anthropologie]: A bit more polished, a bit more older [sic] and she has much less angst.... She tends to be a homeowner and she tends to be in a relationship and more likely than not, married with children.

[Free People]: Independent yet loves being with her friends, her family, and her mate. She travels every spring to festivals, Coachella and Wanderlust being her favorite. She runs and practices yoga to stay fit and balanced. She is influenced by fashion but yet seeks inspiration from all over the world to put together a look that is her own. She is a mix of sweet, cool, and boho and everything in between. We target age 26.[35]

What these categories attempt to capture cannot be defined as a solid object of knowledge; rather, multiple definitions and practices are taken to constitute more-or-less stable "types" through their interactions. The result can best be comprehended as a model, a dynamic object that is constituted from actions that are taken to demarcate predicates. The object of advertising in the control era is here seen to function as a nodal point or black box at which various practice, affects, and essentialized markers of identity intersect and interact.

To be clear, the examples from branding and marketing given here are not intended to stand out as exceptional. They should be taken to represent a general set of ideals and practices that extend across the sociopolitical fabric of the control era. The modeling of social groups according to interactions of discretized predicates (for example, race, age, geography, property, and knowledge) and activities (traveling to Coachella, practicing yoga, "struggling" at art school) emphasizes the logic of definition (and, by extension, of exclusion) that characterizes digital representation. This centrality of modeling to the control-era logic of valorization can be seen to effect a clear division between the set (or set of sets) of traits and behaviors that have already been preempted and included within a given modeling system and the remainder of all the possible traits and behaviors that have not been foreseen and thus to all intents and purposes are denied representation or existence. Indeed, the inclusion of vague categories such as "angst," "sweetness," and "difference" attests to a desire to expand the field of measurement that proliferates categories and congeals affects into fixed objects. In other words, the constant expansion of demographic modeling toward vaguer and vaguer categories points to its grounding in a logical substrate that oscillates between exclusion and universality. It is this division that presents an entry point for socioeconomic critiques, from Deleuze's notion that the dividual replaces the individual in control societies to Tiqqun's claim that "historical conflict no longer opposes two massive molar heaps ... among which, in each individual case, one could differentiate," but that under informatic capitalism "the front line ... now runs through the middle of each of us, between what makes us a *citizen*, our predicates, and all the rest."[36]

As has already been established, the power and diffusion of control's sociocultural logic must be traced not only through digital systems and media but also through nondigital forms whose social conditions of production rely on and thus aesthetically manifest the power of such systems. To simply seek ways in which a given medium directly apes the formal and aesthetic character of digital forms equally fails to attend to this inquiry's basic requirement. Instead, it is necessary to address the question of how one might locate representation of the tension between digital-symbolic depictions of the social and the slow, negotiated struggle (as well as the continuous, rich, and multiple experience) of actual social existence that is of central concern. The analyses presented in the remainder of this book point to solutions to this conundrum, but to properly attend to the relationship between cultural form and cybernetic-capitalist logic it is necessary to address three registers in concert: the technical conditions that provide tools and metaphors for control and the modes of production and distribution pertaining to cultural objects must be examined as inseparable from these objects' aesthetic character.

Beginning from the social and technical principles of modeling, exploitation, and exclusion, the project elaborated here seeks to develop a mode of critical reading attuned to the cultural formations that undergird and sustain the control episteme. This mode of reading allows one to locate the historical emergence of aesthetic formulations of the black-boxed, information-theoretical concepts of the individual and society that are fundamental to control logics of economism and management. Such analyses make the power and diffusion of control visible: if control marks a range of cultural objects beyond computing machines and the practices such machines directly facilitate, it cannot be understood as a source of purely technical relations but instead emerges as the superimposition of these relations, often cunningly shifted into vague metaphors, onto material and social worlds.

4   Text before Images, or The Logic of Digital Worlds

Number series, blueprints, and diagrams never turn back into writing, only into machines.

—Friedrich Kittler, *Gramophone, Film, Typewriter*

Communication, Information, Text

Toward the end of "Postscript on Control Societies," Gilles Deleuze writes that "Kafka, already standing at the point of transition between the two kinds of society, described in *The Trial* their most ominous judicial expressions: *apparent acquittal* (between two confinements) in disciplinary societies, and *endless postponement* in (constantly changing) control societies."[1] Despite his earlier identification of "computers" and "information technologies" as the emblematic media of control, Deleuze's statement that literary production from 1914 expresses the nascent sociopolitical characteristics of this same era serves to reinforce the notion, central to this book, that control societies are above all characterized by a certain historical logic that shapes the conceptualization and management of bodies and minds. As elaborated in the previous chapters, what is essential about Deleuze's "Postscript" is that it locates the base logic of control not in concrete technologies and their uses but in a certain conceptualization of the social as a symbolically representable set of feedback loops subject to statistical modeling. The entirety of social life thus becomes imaginable as a system of perpetual recursion that finds a precursor in the dreams of Babbage and Hollerith, as well as in those of untold bureaucrats.

Why, though, does Deleuze locate the historical roots of control's cultural manifestation in literature rather than in some other form drawn from the range of apparatuses that made up the media ecology of the 1910s—from film and photography to the growing collection of workplace machines and bureaucratic technologies? Where the obvious routes toward aesthetic formulations of control and its attendant tensions would seem to be through the hardware and software of computer media or as theorists such as Lev Manovich have suggested, through screen-based media such as film, television, and video games, Deleuze's specific invocation of Kafka invites media-theoretical questions about how literature might provide specific insights into a digital cultural logic. At the most literal level, one might be guided in responding to these questions by the fact that the etymological root of control, the *contre-rolle*, evokes the practice of writing onto a material substrate such as paper as the root of bureaucratic technologies.[2] Because of its apparent separation from electronic digital computing machines and their applications, Deleuze's specific selection of a literary work as manifesting an early aesthetic trace of control offers a startling provocation from which to begin tracing the emergence of control as cultural logic. This is not to suggest that image-based forms are to be put aside: as both Friedrich Kittler and Sybille Krämer have noted, the difference between analog and digital media can be mapped onto a more fundamental distinction between language and image in suggestive ways,[3] and following this the present chapter addresses the complex relationships between digitality and non-computer-based visual culture. Indeed, the two references to aesthetic objects that Deleuze makes in "Postscript" are to Kafka and television game shows, a fact that sketches out a line along which the cultural valences of control can be seen to diffuse alongside its political-economic adoption.

On the one hand, Kafka's writing is full of machines that gesture toward the regulatory logic of control—from the telephone networks of *The Castle* and "My Neighbor" to the execution machine of "In the Penal Colony," the latter resembling nothing less than a Jacquard loom for inscribing the law directly onto the flesh of the condemned—and figure the superimposition of disciplinary and control apparatuses. This fascination with pre- and protocomputational media assemblages is emphasized in Kafka's letters: Deleuze and Guattari, Kittler, and Bernhard Siegert, for example, have all discussed his letters to Felice Bauer (from 1913) and Milena Jesenská (from 1922), in which Kafka writes first of the prospect of combining analog media technologies such as telephone, parlograph, and gramophone and later of the "ghostliness" of technical communications.[4] But the transition from discipline to control that Deleuze locates in Kafka is centered not only on types of machinery but also on systems of management within which technologies play only one part. Deleuze's interest in this protocybernetic aspect of Kafka's work is evident in the book he wrote with Guattari in 1975 titled *Kafka: Toward a Minor Literature*. Considering the first chapter of *Amerika*, Deleuze and Guattari write that Kafka is exemplary in understanding that

a machine is never simply technical. Quite the contrary, it is technical only as a social machine, taking men and women into its gears, or, rather, having men and women as part of its gears along with things, structures, metals, materials. Even more, Kafka doesn't think only about the conditions of alienated, mechanized labor—he knows all about that in great, intimate detail—but his genius is that he considers men and women to be part of the machine not only in their work but even more so in their adjacent activities, in their leisure, in their loves, in their protestations, in their indignations, and so on.[5]

This identification of a cybernetic dimension to machine-facilitated labor, which presents a grand systematization of labor and nonlabor time based on both technological conditions and metaphors, goes some way toward explaining Deleuze's positioning of Kafka at a pivotal moment in the emergence of control societies. Kafka's work, which predates the electronic digital computer but not those precursor technologies (the machines of Jacquard, Babbage, and Hollerith, for example) that express a desire for a digital worldview fundamentally oriented toward managed and valorized labor, emphasizes the technological production of epistemic formations as well as practices. The endless postponements in *The Trial*, wherein disciplinary confinement alternates with a bureaucratic management of time and space that is permutational and apparently purposeless (to a human observer), express the preconditions for forms of social arrangement that would bring the computer to socioeconomic and cultural ubiquity as both tool and metaphor.

The imperative that Deleuze locates in Kafka suggests that one must examine the relationships between literature and the logical substrate of control systems in ways that go beyond descriptions of past, present, or future technologies and their subjective effects (as in the case of Kafka's bureaucratic apparatuses, Thomas Pynchon's engagement with information theory and duck-mimicking automata, and genres such as the technothriller and cyberpunk, among many other examples). Such representation of existing or imagined machines is, after all, a capacity that is not unique to writing. Moving beyond Deleuze's focus on socioeconomic and juridical systems, literature emerges as a telling representational mode under control in part because writing is reformulated, under the epistemic conditions of cybernetics, into a symbolic system that mirrors the computer's functional logic. In Kittler's work the cybernetic reduction of human affairs to a form of symbolic data processing determines the epoch of modern written discourse. In the introduction to *Gramophone, Film, Type-writer*, for example, he writes three sentences that contain the conceptual foundations of his entire media-theoretical project: "Time determines the limit of all art, which first has to arrest the daily data flow in order to turn it into images or signs. What is called style in art is merely the switchboard of these scannings and selections. That same switchboard also controls those arts that use writing as a serial, that is, temporally transposed data flow. To record the sound sequences of speech, literature has to arrest them in a system of 26 letters, thereby categorically excluding all noise sequences."[6] In Kittler's account of alphabetically coded language, literary aesthetics are nothing more than the product of analog–digital conversion or the capturing of the real by the symbolic. In making this claim, Kittler suggests that the filtering and discretization effected by universal alphabetization for the purpose of literary production is a fundamental step in the emergence of the digital "information channel that can be used for any medium."[7] Yet there is a curious lack of historicism in Kittler's account. The socioeconomic primacy of symbolic logic that marks the emergence of control societies, well under way at the time he was writing his signal books on literature and technical media, is conspicuously absent from his description of the same logic as fundamental to literary production. In other words, Kittler's claim, made in the afterword to *Discourse Networks 1800/1900*, that the "contemporary sociology of literature" fails to address its subject matter because it reads texts "only" as "reflections of relations of production, whose paradigm is energy or labor rather than information," does not account for the historical processes through which information *joins* energy at the base of these relations of production.[8] Where Kittler writes of a technological a priori, then, one can modify his position (against his explicit intentions, to be sure) so that the a priori of the media represents a material historical imposition, namely, the epistemic grounding of control. Making such a move allows one to formulate a rigorous critique of control that is able to take the mutability of capital alongside technological development into account without imposing a technological determinism. In this reconfigured arrangement, literature, once written language has been historically reformatted as a form of digitality, bears witness to the epistemic reconfiguration of human as data processor that in Kittler represents a mediatic a priori and that in this book represents the essential precondition for the control apparatus of post-Fordist labor, neoliberal economism, immaterial labor, and so on.

Drawing on Wlad Godzich, Jonathan Beller suggests that language is supplanted by images through the historical emergence of the attention economy (here understood as one economic valence of control societies) because the slowness of the former is antithetical to the most efficient organization and valorization of attention.[9] This account is compelling, and the cultural dominance of visuality over language that it expresses will emerge as a central element of control culture through this chapter. But it is also problematized by the fact that the technical possibility of the digital computer is bound to a discrete, symbolic register that is (from the viewpoint of computer science and communication engineering) far closer to written text than it is to images, even if dominant cultural practices mask this symbolic register below images and sounds in order to facilitate usability and thus the valorization of user action. "An elementary datum," Kittler writes in the afterword to *Discourse Networks 1800/1900*, "is the fact that literature (whatever else it might mean to readers) processes, stores, and transmits data, and that such operations in the age-old medium of the alphabet have the same technical positivity as they do in computers."[10] In short, from the viewpoint of control, writing and computation resolve to the same logical substrate—the representation of past actions, future possibilities, identity, affects, and objects, as discrete symbols.[11]

From the logical-symbolic formulations set out by Alan Turing and Claude Shannon in the late 1930s to the linguistic formulation of computer code (at the level of abstraction that comes one step above binary code and two steps above voltage differences in physical logic gates) to the options in drop-down menus and checklists that constitute the limits of both identity and possibility in the range of software applications that structure leisure and labor today, a particular deployment of text characterizes many of the principal apparatuses of computer culture. It is thus instructive to locate the ways in which writing can both closely hew to and expose the limitations of the digital worldview that is a precondition for the intensification of capital's cybernetic qualities. The engagement with literature in this chapter, then, is necessary for the simple reason that the symbolic representation and execution of worlds through text is central to the historical, socioeconomic, and cultural aspects of the informatic stage of capital. As Agamben writes, "[t]he writer of stories" is as caught up in the desubjectifying circuits of late capitalism as the "user of cellular phones" and "the web surfer."[12]

Writing and/as Symbolic Representation

The symbolic configuration of the world as network or data processor mirrors the position represented at the Macy conferences by communications engineers such as Shannon. For Shannon, studies of communication should be exclusively concerned with developing methods for distinguishing that which can be subject to predictive analysis from that which cannot. The actual content of a given text is understood not as a singular occurrence but as a selection from a calculable number of alternative possibilities: "semantic aspects of communication," Shannon famously writes in "A Mathematical Theory of Communication," are "irrelevant to the engineering problem." Because of this a workable communication system "must be designed to operate for each possible selection, not just the one that will actually be chosen since this is unknown at the time of design."[13] This cybernetic-linguistic indifference to content and context is emphasized in an exchange between Shannon and the group psychologist Alex Bavelas at the eighth Macy conference, held in March 1951. Rebuffing Bavelas's proposal that information theory might be applicable as a method for analyzing group problem solving, Shannon claimed that "the determination of the semantic question of what to send and to whom to send it" falls outside of the purview of the cybernetic approach to communication.[14] Content and the author–reader relationship, in other words, have no place in the world of probabilistic mathematical communication of which the data-processing power and purely symbolic logic of the electronic digital computer is emblematic. Fascinatingly, one of the text sources Shannon uses to demonstrate this principle is James Joyce's *Finnegan's Wake*, which appears in "A Mathematical Theory of Communication" as an example of a very low redundancy corpus in contrast to the high redundancy of C. K. Ogden's 850-word Basic English vocabulary.[15]

The experiments drawing on Joyce and Ogden as information sources undergird Shannon's subsequent formulation of a language that he calls "Printed English," a system consisting of twenty-seven characters (the twenty-six letters of the Latin alphabet plus a character for white space) to facilitate statistical analysis of the probability that each character would follow a given other. Taking this character set as a basis, Shannon, in "A Mathematical Theory of Communication" and a 1950 paper titled "Prediction and Entropy of Printed English," presents a series of experiments through which the predictability of passages of text can be determined. Under this system, language becomes another component of the probabilistic world, joining the pilot under the duress of attack in Wiener and Bigelow's anti-aircraft prediction, the firing of neurons in a network in McCulloch and Pitts's logical calculus, and the general economic behavior of social actors from the point of view of game theory and automata theory. Just as Kittler suggests, writing, viewed from the point of view of cybernetic principles, becomes nothing more than selection from a field of definable alternatives. That six decades after the last of the Macy conferences literary corpuses are "read" as sources of data—not in communications engineering but under the disciplinary umbrella of literary studies—bears witness to the dissipation of cybernetic logic into the sociocultural fabric of control.

A number of other interfaces between cybernetics and literature point to the confluence of cultural objects and the control worldview. Von Neumann and Morgenstern make reference to *Robinson Crusoe* a number of times in *Theory of Games and Economic Behavior*, although these references generally appear to evoke previous traditions in political economy and are concerned with the broadly defined character of Crusoe rather than with anything specific to Defoe's novel.[16] Later, a passage from Arthur Conan Doyle's story "The Final Problem" in which Holmes must decide whether to stay on or disembark from a train in order to evade Moriarty appears as an example of the zero-sum, two-person game of matching pennies "in an entirely different material garb."[17] Wiener, particularly in *The Human Use of Human Beings*, makes more detailed reference to literary works, including Rudyard Kipling's *Jungle Book* and "With the Night Mail" and Charles Dickens's *Little Dorrit*, in order to illustrate particular scientific and sociopolitical problems that might be approached from a cybernetic perspective.[18] Wiener also published two short stories and a novel in the 1950s. McCulloch and Shannon composed poetry, while Beer included passages written in verse alongside his prose accounts of cybernetics and management. In terms of this historical relationship between literature and information as a paradigm of representation, however, three distinct literary interfaces with the cybernetic reconceptualization of the world are particularly instructive. The first centers on von Neumann's interest in Goethe's *Faust*; the second on an essay on Kipling published by Wiener and Deutsch in 1963; and the third on the contributions of I. A. Richards, the only literary critic to attend any of the Macy conferences, to debates over cybernetics and the humanities.

Accounts of von Neumann's intellectual connection to *Faust* cover nearly the full length of his life, from a childhood fascination with Goethe's play through to his deathbed, where, as William Poundstone recounts, he drew on his renowned eidetic memory to reproduce lines of the text before his brother Michael could read them to him from a printed copy.[19] Moving beyond this use of the text as a source of pure data to which mnemonic capacities can be applied, Chun's analysis of von Neumann's fascination with *Faust* perceptively focuses on the relationship between language and action that it evokes, and in particular on a sentence from part 1 of the play that Nicholas von Neumann suggests was of special interest to his brother: "In the beginning was the act."[20] Although Chun acknowledges Nicholas von Neumann's suggestion that the influence of this passage, along with the later line "the deed is everything, the glory naught," was apparent primarily in his brother's social beliefs about the "redeeming value of practical applications in his profession," she extends a compelling argument about the political logic that is implicit in the Manichean sciences of game theory and cellular automata that the elder von Neumann brother pioneered. As Chun demonstrates, Faust's claim that "in the beginning was the deed" (as the 1987 David Luke translation that Chun cites has it) comes at the end of the monologue in which he grapples, in translating the Bible into German, to satisfactorily come to terms with the claim "in the beginning was the word." Faust's substitution of "deed" for "word," Chun notes, "sums up von Neumann's axiomatic approach to automata and his attraction to McCulloch and Pitts's work. It also leads him to conceive of memory as storage: as a full presence that does not fade, although it may be misplaced."[21] The cybernetic vision of the subject as a symbolic machine, Chun suggests, thus finds a formative influence in a literary reconfiguration of language (as well as of "force" and "mind," the other possibilities Faust considers before settling on *deed* as the cardinal term) as positively measurable (and thus valorizable) action. The human–computer metaphor, the cornerstone of the vision of the social as a network of discrete automata that unfolds in time like a Markov chain, finds a formative influence in a written meditation on writing.

Wiener and Deutsch's 1963 essay "The Lonely Nationalism of Rudyard Kipling" exhibits a mapping of a digital worldview onto the form and content of a literary object similar to that which Chun identifies in von Neumann's fascination with *Faust*. For Wiener and Deutsch the formalization of social interaction into binary poles represents a dangerous tendency, but this does not prevent the authors from identifying it as a universal and naturally occurring worldview rather than one that is produced out of specific sociotechnical conditions. In Kipling, Wiener and Deutsch write, one finds a recurring "horror of ambiguity," a "frantic belief in the all-or-nothing character of group allegiance," and a worldview in which "meetings between different peoples and cultures must lead to sharp 'either or decisions.'"[22] This is a warning about Manichean sociality that maps the binary-switching logic of the computer or the McCulloch-Pitts neuron onto the foundations of community—a fact that literary critic Elliot L. Gilbert did not miss, writing in a 1979 review of Weiner and Deutsch's essay that "[w]hether deliberately or intuitively, the two critics have drawn our attention to what they see as the similarity between the operation of Rudyard Kipling's mind with its binary predilection, and the operation of a computer." Even for Wiener, whose misgivings about the promiscuous application of cybernetics to society are well documented, worlds imaged through literary means are automatically amenable to the imposition of a digital worldview. This cybernetic vision, Gilbert suggests, sheds less light on Kipling's writing than on the dangers of the emerging sociotechnical episteme with whose roots it intersects and in which "an inability to make the crucial distinction between the mechanical and the organic in human structures" poses a fundamental danger.[23]

If von Neumann's reading of *Faust* and Wiener and Deutsch's reading of Kipling demonstrate the habitual imposition of digital logic that is foundational for a cybernetic vision of the social, I. A. Richards's contribution to the eighth Macy conference posits a fundamental opposition between this cybernetic vision of language and the social conditions represented in the possibility of literature. At this conference—the same one at which the exchange between Shannon and Bavelas took place—Richards presented a paper titled "Communication between Men: Meaning of Language."[24] In the published version of this talk, Richards cites Wiener directly and discusses feedback and feed-forward, but he ultimately rejects the notion of digital communication as a basis for social analysis by proposing that attempts to separate linguistics and behaviorism from "other studies" in pursuit of "intellectual world-conquest" were dangerous both for these fields and for others. Running counter to the dismissive accusations of subjectivity leveled at any participant at the Macy conferences who dared to venture into territory not representable in symbolic form even during largely empirical discussions, Richards insisted in his talk that the application of "naïve scientism" to social and cultural fields was a "threat"; that a "'purely descriptive' linguistics" would pose a "grim danger" to "education," "to standards of intelligence," and "to the reserves in theory and in sensibility of the mental tester"; and that the "'scientific objectivity' of which many a linguistic scientist is so charmingly vain ... is out of place when it tries (as it does) to interfere with education or criticism."[25] Although the form of close reading Richards espoused is far more closely aligned (on the political spectrum of the contemporary Anglo-American academy) with positivistic computational methods than those grounded in critical theory and depth modes of interpretation, his intervention at the eighth Macy conference illustrates the historical break that the cybernetic concept of language introduced within the ever-shifting category of literature. Between von Neumann's fascination with language as the rendering of thought into action, Wiener and Deutsch's impression of a computational worldview onto Kipling, and Richards's response to the notion that literature can be reduced to symbolic models, it is possible to trace the contours of debates around digitality and literature that continue today. These claims, each of which was made at a critical moment in the emergence of the control epis-teme, serve as powerful rejoinders to the present cultural manifestations of control among which ahistorical digital literary study can be found.

   Where Kittler studiously avoids the political implications of the exclusion of "noise sequences" within symbolic representation, this chapter argues that such phenomena cannot be considered in any way other than histori-cally and politically because, from the mid-twentieth century onward, logics of filtering and discretization have increasingly been deployed to configure the social as a medium that can be coded and modeled. Literary fiction may formally gesture to the discrete, symbolic register of the computer by way of standardized alphabets and printed writing, but at the same time it ges-tures toward the imaginary—what Margaret Morse calls the "sinking into another world" that links the novel with the theater and cinema and that distinguishes each of these forms from television—in ways that computa-tion does not.[26] Writing, as Krämer argues, represents "a hybrid construct in which the discursive and the iconic intersect"—an observation that even Kittler does not deny, noting as he does that the practice of reading involves "hallucinating a meaning between letters and lines," even if this hallucina-tory potential is diminished by the automation of the means of expression first by technical media for storing text, sound, and images and then by the universal medium of the computer.[27] This tension between a visible sym-bolic register and an invisible imaginary register—an inversion of the cul-tural framing of computation, in which the imaginary of visual culture hides and thus normalizes the function of the discrete state machine—points to the most telling ways in which literature intersects with both representa-tions and critiques of digital logic. Today, in the midst of the society of the networked spectacle, printed writing remains central to engagements with the way informatic rendering shapes social experience.

Figuring the Digital

With these theoretical and historical contexts in mind, one can observe that the relationship between language and action that characterizes cyber-netic logic emerges as the central concern for locating the cultural expres-sion of control in literary production. A model for such an approach can be found in a recent theoretical work on digitality such as McKenzie Wark's book *Gamer Theory.* Here Wark reproduces (while amending the "obsolete" term *cyberspace* to *gamespace*) a statement by Mark Fisher, asking: "'What do we look like from [game]space? What do we look like *to* [game]space? Surely we resemble a Beckettian assemblage of abstracted functions more than we do a holistic organism connected to a great chain of being. As games players, we are merely a set of directional impulses (up, down, left, right); as mobile phone users, we take instructions from recorded, far distant voices; as users of SMS or IM, we exchange a minimalized language often communicating little beyond the fact of communication itself (txts [sic] for nothing).'"[28] This statement is less interesting for its depiction of a social world run through with apparatuses such as video games and mobile phones than for its invocation of Samuel Beckett as an exemplary literary producer of such worlds. This is far from a throwaway association: where Kafka depicts actual bureaucracies as well as the purely linguistic management of life at the boundaries of disciplinary and control societies (the "endless postponement" that Deleuze describes as characteristic of the traces of control visible in *The Trial*), Beckett's formulation of control contains no such explicit references to social management by external agencies and instead takes place almost solely at the level of language—the symbolic-communicative register through which control systems model and manage worlds.[29] Instructive in this regard is Beckett's characterization of himself as an "analyzer, trying to leave out as much as I can," in contrast to James Joyce, who he describes as a "synthesizer, trying to bring in as much as he could."[30] This association of analog synthesis with Joyce and of the range of analyzers in computer history (including Babbage's never-completed analytical engine and the differential analyzer worked on by Vannevar Bush, Shannon, and Wiener at MIT in the 1930s) with Beckett stages the historical tension between digitality and the continuous, excessive character of the material, social, and psychic worlds it models. Although Jacques Derrida appears to contradict this formulation when he describes *Ulysses* and *Finnegan's Wake* as computers in "Two Words for Joyce," his account can serve to evoke the difference between the actually existing computer, whose metaphorical extensions to people and sociality always exclude, and the idealized computer of control, which promises to model the fullness of all existence and possibility without exclusion even as if effects the types of exclusion detailed above. The Joycean computer, Derrida states, is a "1000th generation computer ... beside which the current technology of our computers and our microcomputerified archives and our translating machines remains a *bricolage* of a prehistoric child's toys." For Derrida, the computer capable of integrating "all of the variables, all of the quantitative or qualitative factors," that mark the interconnections of "phonemes, semes, mythemes, etc." that constitute the novel as event (as opposed to simply measuring the redundancy of the words that make up the text, as in Shannon's study) would produce the "double or the simulation of the event 'Joyce,'"[31] which is to say, a double of the entirety of nature (to use Kittler's term for the totality that cannot be fully computed by any actually existing digital system).

Beginning from Deleuze's claim that cultural forms and technologies develop as parallel expressions of emerging social logics, the remainder of this chapter progresses by tracing historical parallels between Beckett's writing, the emergence of the control episteme, and the development of the digital computer as a technical and cultural form from the 1930s to the 1980s. This is not an archival or genetic project, looking to find material connections of influence between the electronic digital computer and Beckett's writing (although one might locate at least the germ of such a link in the enthusiasm for Leibniz that he shares with cyberneticians such as Wiener[32]). Rather, Beckett's writing and the emergence of multimedia, user-centered computing are here read together for traces of the historical conditions that shaped and were shaped by the emergence of the computer as tool and socioeconomic metaphor. If, as Marx writes, "the forming of the five senses is a labour of the entire history of the world down to the present,"[33] then it ought to be possible to locate cultural traces of the emerging control episteme across both aesthetic production and technical research and development, both of which are produced under the same social, political-economic, and technological conditions.

The engagement with Beckett's work in this chapter, then, is an exercise in comparative cultural analysis that examines the shared formal components of two systems, one literary and one technical, in order to create an indicative picture of the cultural-political terms that define the present age of control as ubiquitous cybernetic logic. In order to pursue such a project, a degree of formalism is required even if it is (thankfully) destined to ultimately fall short of a full systematization of the subjects in question. The striking parallels between Manovich's tripartite formalization of new media technologies and the three progressive languages that Deleuze identifies in Beckett present a compelling framework in this regard. In *The Language of New Media*, Manovich sets out a three-layered model for the analysis of digital media, stating that the new media object is "digital on the level of its material," "computational (i.e. software driven) in its logic," and "cinematographic in its appearance."[34] With arguments about the limitations of such a rigid formal model put aside for now, a historically tiered deployment of this structure serves as a useful framework for the analysis of the electronic digital computer's technical and cultural emergence to ubiquity, under which the exclusionary logical substrate of binary formalization is progressively hidden behind user-friendly programming languages, operating systems, and images.

Parallel to the historical development of Manovich's three layers of computation, the three languages Deleuze locates in Beckett affect a similar obfuscation of discrete symbols by combinatorial arrangements, sound, and images. Deleuze identifies these languages in an essay titled "The Exhausted," published six years after *Foucault* and two years after the "Postscript." In "language I," which finds its definitive example in Beckett's novel *Watt*, "enumeration replaces propositions" and "combinatorial elements replace syntactic relations." "Language II" emerges through the trilogy of novels *Molloy, Malone Dies*, and *The Unnamable* and finally "blares forth" from the radio pieces. It differs from language I in that it "no longer operates with combinable atoms but with bendable flows." Finally, "language III" develops from *How It Is* in 1961 to a point in the early 1980s where it finds the "secret of its assemblage in television." It is "no longer a language of names or voices, but a language of images."[35]

If one superimposes the historical periods delineated by Deleuze's study of Beckett onto Manovich's layers of new media, it becomes clear that the first of Deleuze's periods, language I, corresponds historically and formally to the early theoretical and technical developments of digital computation carried out by Turing and Shannon in the late 1930s and is concerned with the symbolic formalization of experience into algorithms ("enumeration" and "combination" rather than propositions and syntactical relations). The period in which language II emerges corresponds to the emergence of programming languages and techniques for analog–digital conversion that introduced human-centered layers over the bare algorithmic processes of computing machinery. Through these developments, computation begins to appear, however illusorily, as a kind of "bendable flow"—something that can be used for a range of applications and that is thus directed toward social and cultural applications—rather than as a string of "combinable atoms" whose mechanical abstraction from concrete experience is self-evident. The final language, language III, corresponds to the development of computer graphics in the late 1960s and gestures toward the sociocultural construction of the computer as multimedia machine that Kittler describes in the introduction to *Gramophone, Film, Typewriter*.[36] In each instance, a historical-formal parallel with one of the three layers of digital media set out by Manovich is clear. Each language presents an aesthetic counterpart to a stage in the progressive technical development of user-centered digital systems that parallels (but does not determine) the emergence of control societies. In so doing, these languages register a historical logic of control that exists beyond the computer itself and that shapes concepts of perception and possibility through the strictures of digital logic.

Language I: Algorithmic Rendering

Watt was first published in 1953, but Beckett began writing it twelve years earlier.[37] Hugh Kenner, in an examination of the text's "Cartesian sentences" in his book *The Mechanic Muse*, makes the observation that in the formalizing processes of the novel's central character "we're close to the languages of digital computers, which weren't heard of till a decade after *Watt* was written."[38] Kenner's somewhat general statement ignores the mathematical and technical fundamentals of digital computation set out in Turing's 1936 paper "On Computable Numbers, with an Application to the *Entscheidungsproblem*" and Shannon's 1937 master's thesis, "A Symbolic Analysis of Relay and Switching Circuits," although he does acknowledge George Boole's algebra, which grounds both works. By the time *Watt* was completed, Turing's and Shannon's work had intersected in the construction of cryptographic methods and code-breaking machines during World War II, and von Neumann's major elaborations of game theory and the computer architecture that would bear his name had been completed. *Watt*, then, was written in the middle of one of the most productive moments of conceptualization and elaboration in the history of control societies—the moment in which the technical foundations of electronic digital computing were established through the procedure of abstracting a series of "on" and "off" states from a range of logic problems and the development of techniques for implementing these problems in physical switching circuits.

The coding and patterning processes that Watt engages in at Mr. Knott's house are uncannily commensurate with the early developments in electronic digital computation that the novel historically parallels. Watt processes events in the world by branching through every possibility in a formalization of experienced events and alternate possibilities. Kenner formalizes this connection in *The Mechanic Muse* when he formats a paragraph of text concerning Mrs. Gorman's visits first as code approximating the Pascal programming language and then as a series of conditional statements.[39] The paragraph in question appears in Beckett's prose as: "Mrs. Gorman called every Thursday, except when she was indisposed. Then she did not call, but stayed at home, in bed, or in a comfortable chair, before the fire if the weather was cold, and by the open window if the weather was warm, and if the weather was neither cold nor warm, by the closed window or before the empty heath."[40] Kenner gives the Pascal-approximating formulation of this paragraph as

```
PROGRAM MrsGorman (Input, Output);
CONST
   Indifferent = 60
VAR
   Thursday, Indisposed, Called: BOOLEAN;
   Bed, Chair, Hearth, Fire, Window, open: BOOLEAN;
   Rand, Temperature: INTEGER;
BEGIN {Main Program}
IF Thursday THEN
   IF NOT (Indisposed)
      THEN Called:= True
   ELSE {If Indisposed}
      Called:= False;
   IF NOT Called THEN Random;
      IF Rand = 0 THEN (Bed)
         ELSE {if Rand = 1then}
         BEGIN {Else}
            IF Temperature < Indifferent
               THEN (Chair and Hearth AND Fire)
            ELSE IF Temperature > Indifferent
               THEN (Chair AND Window AND Open)
            ELSE IF Temperature = Indifferent THEN
               BEGIN {Else if}
                  Random
                  IF Rand = 0
                     THEN (Chair and Window AND NOT Open)
                  ELSE {if Rand = 1then}
                     (Chair AND Heath AND NOT Fire)
            END {Else if}
   END {Else}
END {Main Program}41
```

Kenner then renders the paragraph as a natural-language algorithm:

Mrs. Gorman
  came, yes/no
  didn't come, yes/no;
if she didn't come then she stayed home:
  in bed, yes/no
  in chair, yes/no;
if in chair, then
  by hearth, yes/no
  by window, yes/no;
if by hearth, then fire burning, yes/no;
if by window, then window open, yes/no.[42]

These expressions emphasize the proximity between the processing of events and possibilities in *Watt* and in digital machine computation. The passage reformulated by Kenner is far from an isolated example of such exhaustive formalization in Beckett's novel; the second part of *Watt* contains a large number of such passages, with the one about Mrs. Gorman's visits representing a relatively short example. In an analysis that emphasizes the proximity of these passages to logical formalization, S. E. Gontarski and Chris Ackerley observe that the manuscripts for *Watt* contain "exhaustive truth-tables" with each permutation ticked off "in such as way as to cover every possibility."[43] Watt's formalization of Mr. Knott's meal arrangements and the days on which leftovers should be given to a dog is among the most extensive of these algorithmic renderings, taking up fourteen pages of the novel and including two passages (the composition, instigation, and execution of Mr. Knott's arrangements as well as the identity, selection, and ownership of the dog) that are comparable to subroutines in programming.[44] In each of these examples, the specificity of the actually occurring event is subordinated to the field of possibilities constituted by the sum of included variables (how much of Mr. Knott's meal is left over, the possibilities for locating an appropriate dog, and the bringing together of food and dog in time and space)—an arrangement that evokes Wiener and Bigelow's predictive modeling and Shannon's mathematical communication as well as the organizational and economic practices that follow them.

If the second section of *Watt* is concerned with representing the imbrications of subjectivity and algorithmic formalization that mark the emergence of control's human–computer conflation, the third section implements the method elaborated in Shannon's paper "A Mathematical Theory of Communication" by framing communication as a statistical system. In this section, Watt begins to reverse the order of the words in his sentences, then the letters in his words, then the sentences in his statements, then both the words in the sentence *and* the letters in the word, and so on. From the perspective of Shannon's theory, these modes of communication are functionally identical because they are patterned in a way that enables their reliable reconstruction by statistical process. As Shannon notes in "Prediction and Entropy of Printed English," when test subjects were given the last letter of a word and asked to guess its predecessors, "the scores were only slightly poorer" than when they were asked to work forward through the sentences, even though this reversed version of the test was "subjectively much more difficult."[45] There is not much in the way of sophisticated cryptography here.[46] The idea of intelligibility to a human reader or listener is not a consideration in Shannon's communication theory; all that matters is the optimization of signal strength and the likelihood of a message being efficiently reconstructed by the receiver. From a technical (rather than an interpretive) perspective, the patterns of speech in the third section of *Watt* detail not a deterioration of Watt's mental state but rather the configuration of an output stage that corresponds to the method for processing input demonstrated in the algorithmic formalizations of the previous section. In this section—the part of the novel that accounts for the final period covered by its narrative—Watt adopts a mode of output that is commensurate with the mode of analysis he had already perfected. Communication is here optimized for parsing rather than for reading.

As both Turing and Shannon demonstrated in the years preceding the composition of *Watt*, anything can be represented in switching circuits provided it can first be formalized as a finite number of discrete units. As Kittler shows, this historical possibility has significant implications for things that cannot be represented in this way: "Weil nur ist, was schaltbar ist" (only what can be configured as a switching circuit exists).[47] Beckett, though, does not present this principle without its limitations and inherent traumas. Occasions that exceed finite formalization trouble Watt and serve to draw a distinction between perfectly codable phenomena and those that resist configurations appropriate for switching circuits. Examples of such phenomena appear in the early stages of *Watt* in the form of a song whose two verses recount two recurring decimal numbers, 52.285714 and 52.142857. These numbers, corresponding to the simple calculation of the number of weeks in a leap and in a regular year respectively, are infinitely recurring and therefore pose a problem to representational procedures that require finitude.[48] For Watt, as Kenner observes, imprecisely definable conditions of possibility provide the experiential equivalent of these numbers later in the novel: how Mrs. Gorman *feels*, for example, when in her bed or chair, in front of window or hearth, and so on.[49] The parallels with the black box and W. Ross Ashby's insistence on the methodological necessity of excluding consciousness from his cybernetic modeling of the brain scarcely need elaboration here. The tension between the configuration of experience as algorithm and the exclusion of material that exceeds formalization in Beckett points to the tensions that inhere in the cultural traces of control that parallel (rather than trail) the emergence of computer culture. Watt, whose namesake produced the steam-engine governor that would provide cybernetics with a conceptual model of self-regulation, is characterized as an overly literal version of the human computer that represents the grounding image of control. As Wendy Brown notes of neoliberalism, the "correct" internalization of the values inherent to such a system requires systems of education and normalization in order to function as intended, and Watt fails to adopt these norms even as he internalizes the cybernetic worldview. The third section of the novel crucially reveals that Watt is recounting his experiences in an asylum: like the tramp who in the film *Modern Times* (Charlie Chaplin, 1936) finds himself institutionalized for internalizing the logic of Fordist production too fully—that is, in a way that makes him an impediment to the workings of the factory—*Watt* diagrams the line between "normal" and "pathological" adoptions of the machinic metaphors that constitute the control episteme. This line, *Watt* demonstrates, is constituted by the social actor's capacity to be productive rather than purely analytical and recursive in their capture, processing, and communication of data.

Language II: Executable Text

Although the algorithmic renderings carried out by Watt—like Turing's resolution of logic problems to computable numbers, Wiener's materially limited cybernetics, and Shannon's information theory—exemplify the digital modeling of worlds that is an essential component of control, they are antithetical to value-productive work. As Kenner writes, even when formalized into Pascal, Watt's analyses of Mrs. Gorman's visits and Mr. Knott's meal arrangements "don't give the computer anything to do."[50] It is telling, then, that the body of writing that constitutes language II begins with a passage from exactitude to flexibility that parallels the historical emergence of control-era labor models to supplement the disciplinary principles of constrained, linear, clock-bound work. In the second part of *Molloy* (1951), the detective Moran, whose domestic life is characterized by the fastidious procedures of time management and order of the ideal Fordist subject, is charged with a job that dissolves the order and rigidity of these practices from the moment it is proposed. "I remember the day I received the order to see about Molloy," Moran writes. "It was a Sunday in summer. I was sitting in my little garden, in a wicker chair, a black book closed on my knees. It must have been about eleven o'clock, still too early for church."[51] It is at the moment that Moran is put to work—that is, the moment that he learns of the existence of Molloy, whom he must somehow locate—that his procedural method begins to break down. Moran states soon after his meeting with Gaber, the agent who charges him with locating Molloy, that the job seems "unworthy" of him and that he cannot take it seriously. In this moment the imperative of *paidia*, or unstructured play, appears incommensurable with the Protestant ethic and the Fordist organization of work. Through the remainder of the novel, Moran attempts, with decreasing traction, to assert order over the task with which he is faced. When considering the "Molloy question," Moran must constantly remind himself of the virtues of classical rationality, defining his "methodical mind" and stating that a "prolonged reflection as to the best way of setting out" always forms "the first problem to solve, at the outset of each inquiry."[52] Yet as Moran prepares to set off on his ill-defined quest to locate Molloy, he finds his systematic, imperative mode troubled by the possibility of unconsidered alternatives and imperfections: "[T]o my son I gave precise instructions. But were they the right ones? Would they stand the test of second thoughts? Would I not be impelled, in a short time, to cancel them? I who never changed my mind before my son. *The worst was to be feared*."[53] Whereas the first part of the novel contains the oft-described permutations of sucking stones that Deleuze describes as an example of the "inventory of peculiarities pursued with fatigue and passion by larval subjects"[54]—the type of formalizing analysis that, although digital and combinatorial in character, appears pathological from the point of view of control because it is nonproductive—the second part emphasizes the conditions of emergence upon which such formalization must be implemented in order to become productive. Language II thus appears out of the formalizing horizon of "language I" in a trauma of emergence, manifesting the same intractability between linear rationality and complex systems that drove the development of cybernetic methods during World War II and sustained these methods into an ever-widening field of social, political-economic, and cultural applications into the twenty-first century.

This language II, the language of "bendable flows," requires a turning of the precision of discrete units (be they identical units of clock time or the "combinable atoms" of symbols that can be configured as sequences of switchable states) toward action rather than (or as well as) toward modeling and analysis. Language II thus mirrors the function of programming languages and software, which were developed across the same historical period. Programming languages instrumentalize the binary logic of the computer, allowing for a mediation of human desire and machine function. They represent processes of abstraction ordered from the machine's perspective: in the same way that the most familiar of home movies viewed as a stream of binary 1s and 0s would appear at a high level of abstraction to a human user, a "language" consisting of anything other than binary electrical variations is at one or more levels of abstraction from the computing machine's perspective. This break between human and machine is emphasized by the layers of translation that exist between different levels of programming language and the hardware of logic gates: a program written in a high-level language such as C, the same program written in a low-level language such as x86 Assembly, and the same program again written in binary code all resolve to identical hardware operations, regardless of their user friendliness. The difference between these levels of language is trivial for the machine; it is nontrivial only for the user.[55] All programming languages are in effect abstractions of the computer's technical function, grounded in switching circuits, in order to make that function accessible and thus conducive to productivity.[56] At the same time, by obfuscating the machine's bare digitality while retaining its logical structure, such languages, like even higher-level abstractions such as operating systems and video games, allow for a commingling of human and digital system that disperses and normalizes the fundamental tenets of control. In the texts of language II it is possible to observe the constitution of action from formal logic that mirrors the instrumentalization of the computing machine and the real subsumption of the (selected, optimized) social actor.

Kenner's examination of Beckett's play *Endgame* (1957) is instructive when tracking the historical movement from language I ("atoms") to language II ("bendable flows"). Written in the period between 1952 and 1956—five years after Herman H. Goldstine and von Neumann completed their report *Planning and Coding of Problems for an Electronic Computing Instrument*[57] and roughly the same period in which FORTRAN was being developed by IBM—the example Kenner cites concerns the exact positioning of Hamm's chair by Clov:[58]

HAMM: Put me right in the center!

CLOV: I'll go and get the tape.

HAMM: Roughly! Roughly! [Clov *moves the chair slightly*] Bang in the centre!

CLOV: There!

   [*Pause.*]

HAMM: I feel a little too far to the left. [Clov *moves the chair slightly.*] Now I feel a little too far to the right. [Clov *moves the chair slightly.*] I feel a little too far forward. [Clov *moves chair slightly.*] Now I feel a little too far back.[Clov *moves the chair slightly.*] Don't stay there [*i.e. behind the chair.*][59]

Here, while high-level programming languages emerge in the labs of IBM, the machine-readable formalizations that don't "give the computer anything to do" in *Watt* are replaced with an imperative language that aims to directly produce action out of natural-language communication. This growing human orientation is the source of Kittler's critique in "There Is No Software," in which he argues that software obfuscates the logic of the machine in order to facilitate human-centered instrumentality, and in so doing tragically prohibits access to that which is specific to the computer medium (if it can be thought of as a medium, as opposed to the end of media). Although Kittler's critique is ostensibly directed at nostalgic humanists and monopolistic software developers, it is not difficult to shift it toward digital modes of political and economic governance: it is not the case that software is fundamentally less algorithmic or nonhuman than hardware, but rather that software masks these characteristics in order to facilitate the precise imbrication of human and computer behaviors that is both materially and metaphorically essential to the emergence of new forms of organization and valorization.

In issuing orders the way Hamm does, Kenner writes, one comes "even closer [than Watt] to the spirit of programming languages, FORTRAN and Pascal and their many siblings, since they are unique in having but one mood, the imperative."[60] In the passage of *Endgame* reproduced earlier, instructions are given for the procedural movement of an object in a way that precludes any of the direct contact with data and algorithms that might be thought of as madness. In an echo of Watt's institutionalization for prefiguring Shannon's mathematical theory of communication, Kittler critiques the ideological connection between insanity and an interest in hardware-level programming in his essay "Protected Mode." Citing a trade publication's claim that "even under the best circumstances, one would quickly go crazy from programming in machine language," Kittler writes

"At the risk of having gone crazy long ago, the only thing one can deduce from all this is that software has obviously gained in user-friendliness as it more closely approaches the cryptological ideal of the one-way [i.e. irreversible] function."[61] As Deleuze and Guattari so famously wrote of the schizophrenic, historically contingent definitions of insanity can reveal much about the logic of capital in a given historical moment. The further one goes toward high-level programming and software applications, the further one moves from the machine's technical function and the possibility of appearing to be an aberrant subject from the perspective of capital. In the same way, the further into language II one looks, the further one moves toward the instrumentalization of Watt's "crazy" algorithmic processes and the closer one comes to the "correct" internalization of digital logic that distinguishes the ideal citizen in the age of control societies. At the same time, the problems of positioning Hamm's chair in the sequence from *Endgame* indicate the kinds of imprecision that can emerge from the relation between algorithmic processes and those concrete social situations that can be neither coded nor eradicated.

Alongside the emergence of programming languages, a second major development in the user orientation of information technologies doubles the language II period in Beckett in the form of methods for digitizing analog inputs such as voices, nonverbal sounds, and images. Deleuze does, after all, observe that language II culminates (which is to say that it begins to transition into the image-based language III) not in writing but in the "blaring" sound of the radio. The first of the radio pieces, *Embers*, was completed in early 1959, placing it at a moment in which the techniques for PCM, under development since the late 1930s, had become well established and work on the Fast Fourier Transform (FFT) that would enable the digitization and microlevel analysis of analog signals for theoretically optimal noise reduction was close to applicability.[62] Alongside the emergence of the conditions of possibility for the computer as multimedia machine, then, Beckett begins to work primarily with technically transmitted voices instead of with text. And after his writing moves from the formalization of data and algorithms to the execution of actions and the subsumption of the analog under the digital, it becomes possible to observe the emergence of language III, the language of images that coincides with the emergence of graphical and ubiquitous computing.

Language III: Executed Image, Executable Subject

As Deleuze specifies, language III begins with the novel *How It Is* and culminates with Beckett's television plays. The period that constitutes language

III thus spans the technical development of graphical computing, from the earliest experiments with vector displays to the array of graphical software interfaces that are today synonymous with personal computing in all of its many forms. Beckett completed *How It Is* in 1961, the same year that Ivan Edward Sutherland began work on his Sketchpad system at MIT's Lincoln Lab.[63] In 1963, the year in which Sutherland submitted the doctoral thesis based on Sketchpad, Beckett wrote the screenplay for *Film,* followed by *Eh Joe,* the first of the television pieces. By the time of *Ghost Trio* (1975) and *... But the Clouds...* (1976), the Xerox Alto Personal Computer had been developed, and with it the first GUI and the desktop metaphor. The period from the late 1970s to the early 1980s saw the emergence of the commercially available personal computer and the associated graphical software packages, including the Xerox Star operating system and the Macintosh 128k operating system. The production of the television play *Quad* (1981) also coincides with the development of the Intel 80286 microprocessor that motivated Kittler's critique of hardware-obfuscating techniques in "Protected Mode."

At the very start of the GUI era, Sutherland's Sketchpad removes not only the need for the user to program in low-level or machine language but also the apparent need to program at all in the sense of inputting discrete symbols (be they the presence or absence of holes in a punch card or sequences of letters entered on a keyboard). Presenting an interface based in part on dragging vectors directly onto a screen, the system combines the analog input of human gestures with the analog visual output of continuous lines. If language I coincides with the theoretical and technical possibility of digital computation and language II with the instrumentalization of computing through the first high-level programming languages, the emergence of software, and the possibility of analog-to-digital conversion, then language III coincides with the emergence of interfaces that both motivate user action and allow it to be captured and algorithmically expressed without requiring the user to enter language of any type. Equally, if the passage about Mrs. Gorman's visits quoted from *Watt* and its corresponding stage of computer history are concerned with the discretization of the real, and the passage from *Endgame* and *its* corresponding stage of computer history are concerned with the abstraction of pure algorithmic logic through imperative commands that resemble natural languages but that directly produce action, then Beckett's work after *How It Is* can be seen to remove the symbolic trace of the algorithm in order to present input and output as simultaneous and inseparable. The exemplary compression of formalization and visuality in this final language can be found in *Quad.*

In *Quad,* four bodies enact an exhaustive series of combinations, each triggering a specific lighting scheme, percussion instrument, passage of movement, and footstep sound. The play presents a defined set of possibilities executed procedurally and expressed visually, and because of this marks both the purest expression of language III and the strongest point of connection between Beckett's writing and the emerging sociocultural conditions of control societies. The piece is filmed with a single static camera located slightly above the depicted space and consists of four hooded figures (figure one wearing white, two yellow, three blue, and four red—they are named for these colors in the script) executing a series of movements within a square "six paces" in length.[64] Each player works through a predetermined course, and each course is based on the lines between the four corners of the quadrangle and the necessary evasion of an invisible square set at the center. The stage directions for this movement consist solely of the letters given to each corner—A, B, C, and D—and appear as follows:

Course 1: AC, CB, BA, AD, DB, BC, CD, DA
Course 2: BA, AD, DB, BC, CD, DA, AC, CB
Course 3: CD, DA, AC, CB, BA, AD, DB, BC
Course 4: DB, BC, CD, DA, AC, CB, BA, AD[65]

Through the execution of these series four times, with a player added at each repetition, every possibility of each course with each combination of players is worked through sequentially. The piece as a whole presents a gridded configuration of space and a digital system of movement that nonetheless produces a visual layer whose effect is to mask the presence of the underlying digital logic. Despite this obfuscating visual layer, the players' movements remain limited by a control-based vision of possibility whereby restrictions are enforced not by a human agent (such as a boss, police officer, or umpire) but by an internalized prohibition on deviating from a coded path.[66]

Indeed, the system of movement that comprises *Quad* appears to be representable as Gray code—a type of reflected binary code posited in a 1947 patent filed by the Bell Labs engineer Frank Gray as a way to minimize errors when implementing binary logic in material switching circuits.[67] Unlike regular binary encoding, successive Gray codes in a sequential list change only by a single digit, so that aberrant outputs resulting from the imperfect synchronization of multiple switches are minimized or eliminated altogether. The relationship between these two coding systems, one implemented in physical switching circuits and the other in social reality through human bodies, is only underscored by the existence of a class of such codes named for both Gray and Beckett—Beckett-Gray codes—which are able to represent the kind of first-in/first-out sequence enacted in *Quad*. The television play, then, imposes onto the players' bodies a logical substrate appropriate for representation in a switching circuit, and produces electronic images of the outcome. It is a tantalizing coincidence that before filing his 1947 patent Gray worked on scanning systems for television and contributed to the foundational paper on raster scanning.[68]

On the one hand, then, the formal system manifested by *Quad*—comprising the digital coding of all possibility, the execution of this possibility in physical space, and the transmission of the overall process for visualization on an electronic screen—is emblematic of the cultural logic of control. On the other hand, however, a glance at the minimal script for the piece foregrounds the impossibility of perfectly coded execution. The script for *Quad* notes that allowances must be made for "time lost at corners and centre" as well as for the "rupture of rhythm" caused by "three of four players" crossing paths at the center.[69] This piece, it is made clear, might reduce perfectly to its coded sequences on paper or as a mathematical model, but not when performed by human actors. That these "allowances" are not prescribed in the script but must instead be devised by the specific director and actors staging and performing the piece emphasizes both the incorporation of individual and group problem solving within the post-Fordist regime and the necessary presence of unforeseen and unmodeled (or unmodelable) behaviors within it. Remarkably, this incommensurability is extended to the level of formal logic when one notes that the formulation of the movements of *Quad* in Gray code is actually impossible when one attempts to implement the rule that the player who leaves after a given cycle is always the one who has been on stage longest. No Beckett–Gray code has been found for a sequence in which n (the number of separate objects or actors in the sequence) = 4 (the prescribed number of players in *Quad*). From the point of view of specifically hardware-oriented binary coding, then, *Quad* in fact depicts an aberrant system that (under current conditions of knowledge, at least) eludes the most error-averse method of implementation as a switching circuit even as it appears perfectly arranged for such implementation.

Think back to Fisher's account (via Wark) of the Beckettian nature of digital social logic, which reduces all possibility to discrete directional impulses and all communication to the *langue* of symbolic exchange. That the privileging of "abstracted functions" over "being"—or an elevation of that which has been parsed and represented as a switching circuit—is diagrammed across the three languages of Beckett's late work is suggestive of a broad series of sociocultural transformations that are not the same as the development and rise to ubiquity of the computer but that are nonetheless linked to this universal, emblematic technology through a shared historical logic. *Quad*, exemplary of language III, draws together the features that develop through languages I and II and that parallel Manovich's three elements of computation. The algorithmic formalization of all possibility; the instrumentalization of computers through the obfuscation of their barest digital logic under natural-language vocabularies, user-friendly syntax, and the incorporation of quasi-analog flows; and the completion of this instrumentalization and obfuscation of the algorithm through images: these three elements constitute the predominant technical-cultural concerns of the present, and the examination of their emergence represents an essential heuristic procedure for the analysis of control's aesthetic manifestations.

But formalization and execution in Beckett always exist in the presence of noise, if one takes this term to describe those emergent elements of being or experience that cannot be captured, modeled, represented, or made executable under the finite conditions of cybernetic systems. The experiences that trouble Watt—such as the recurring decimals produced by attempts to calculate the exact number of weeks in a year and the impossibility of addressing the way Mrs. Gorman might be feeling—are examples of such noise, as are the failures to perfectly execute directional commands in *Endgame*, Moran's loss of control over his discursive order in *Molloy*, and the problems of negotiating the center and formalizing the first-in/first-out queue in *Quad*. This is an aspect of Beckett's writing that intensifies across the three languages. In defining language III, Deleuze identifies not only an abstraction of image from text but also the persistence of pockets of "silence," or absent information, within regimes of images. Language III relates no longer to "enumerable or combinable objects" or to "transmitting voices," but to images with "immanent limits that are ceaselessly displaced—hiatuses, holes or tears that we would never notice."[70] Beckett's writing, for Deleuze, is concerned with the parallel unfolding of two meanings of exhaustion, one related to absolute formalization and the other to the ruptures and supplements that foreground the impossibility of such formalization: "The greatest exactitude and the most extreme dissolution; the indefinite exchange of mathematical formulations and the pursuit of the formless or the unformulated. These are the two meanings of exhaustion."[71] If Kafka's writing describes a specific thread of the intertwined historical processes that constitute control—that is, the emergence of perpetual modulations beyond the fixed boundaries of disciplinary sites—it is possible to identify a dual movement across Beckett's writing: on one hand, the three languages depict increasingly normalized processes of abstraction and the representation of human activities as discrete, valorizable units; on the other hand, they retain the presence of objects that cannot be rendered in this way, or that remain stubbornly resistant to digitization.[72]

As Tiqqun put it, for all of the *"universal enrolment"* and *"proliferating schematization"* that social formations grounded in the control episteme strive toward, the fact remains that "[e]entropy, considered as a natural law"—that is, as a general reminder of the impossibility of absolute formalization—"is a cybernetician's hell."[73] For all of the logical abstractions that Beckett's work presents, markers of those things that escape the world picture of control remain ever present. This is not to say that the moments at which these anomalous phenomena appear are joyful. On the contrary, they are almost always associated with uncertainty and discomfort. They are moments that tend to result in expulsion. Watt wanders the streets, finds insecure work as a manservant, and is incarcerated in an asylum. Moran's orderly life crumbles and he becomes a double of the vagrant Molloy, culminating in the revelation that the novel as a whole is running in a loop. The walkers in *Quad* loop through their patterns with the anxiety of breakdown or emergence recurring at the center zone. Therein lies the ambivalence of unmeasurability in the age of control societies. It should be remembered that the impossibility of digital representation does not always lead to the heaven of escape (or the cybernetician's hell). As the preceding chapters show, exclusion is methodologically built into cybernetics and will remain a principal source of dispossession while the dominant logic of global management remains that of control. To be excluded from the world picture (or world model) that control facilitates is to fall into the realm of precarity at best, and the realm of surplus or remaindered life at worst. To find ways in which to evade or impede this world model without foreclosing on one's prospects for survival, and to have nothing to escape from anymore: this is the ideal toward which metaphorical evocations of noise and entropy only gesture.

5   From Narrated Subjects to Programmable Objects

In warding off the "hell" of entropy, the symbolic capture (or the elimination) of the subject is essential for the sociocultural emergence of control. This is clear in Dupuy's observation that both cybernetics and neoliberal economism are characterized by the conceptualization of social actors as "subjectless processes" and in Agamben's account of media as affecting processes of desubjectification under late capitalism.[1] Beyond these theoretical accounts, the same process is visible in the historical reformulation of sociality, communication, and affect as labor and the extreme flexibility/precarity of disposable labor across the Global South and in the fissures of the Global North, all of which are supported by sophisticated systems of rankings and metrics that obfuscate the material and experiential conditions of the actual worker. This chapter turns to the conceptual and aesthetic construction of the symbolically reformulated subject as norm through historical analyses first of psychoanalysis and then of narrative production, the "socially symbolic act" that Jameson identifies as a privileged space for the imaginary or aesthetic working through of otherwise irresolvable historical problems.[2]

One valence of the relationship between digitality, subjectivity, and narrative can be explicated in terms of memory. As Warren Sack has observed, there is a clear distinction between the notion of memory that is central to modernist aesthetic practices (and, by extension, that which recurs in classical forms of cinema) and the notion of memory that is articulated in computer science. Memory, in the latter case, no longer takes a form analogous to "Marcel Proust's lost aristocratic memories of, for instance, eating scallop-shell shaped, lemon-and-butter flavored cakes as a child." Instead, computer science literature is "filled with the memories of bureaucrats: numbers, lists, tables, cells and segments. *Even the computer science literature on narrative memories boils down to a set of techniques for fitting stereotypical stories into preconceived grids.*"[3] Although Sack makes reference to the

DARPA-run Message Understanding conferences of 1987–1997, one does as
well to examine a much earlier example concerned with the application of
procedures of sampling and gridding to folktales, the privileged subject of
structuralist narratology. In a 1963 article for the *Journal of American Folklore*, B. N. Colby, George A. Collier, and Susan K. Postal describe the computational analysis of forty-five thousand words' worth of folktales drawn
from five cultures in an attempt to locate distinctions and similarities based
on recurring formal properties. Configured for an IBM 7090 computer, the
task necessitated a foundational practice of exclusion, removing both "high
frequency words of low information content" and "words of medium frequency which have multiple meanings and are therefore ambiguous out of
context" in order to focus on "words with single predominating meanings
which are classified under one of 180 themes." This process is premised on
abstraction and noise reduction; Colby, Collier, and Postal are quite explicit
in stating that "the eliminating of high frequency words and those with
multiple meanings is a regrettable information loss but it is in the interest
of greater analytical clarity and accuracy."[4] Although both of these accounts
of the digitization of stories are instructive for the analysis of the cultural
logic of control, they are focused on a particular procedure—namely, the
digitization of already-existing narratives. This chapter is concerned with
a quite different set of historical processes: those through which a certain
concept of the subject emerges, functions within the socioeconomic order,
and is thus manifested in new narrative forms that are produced under the
emergent epistemic conditions of control. This computational concept of
the subject, which properly ought to be defined as a "subjectless process" or
a "desubject," can be understood as digital in general and as programmable
in particular. Because of this it is necessary to spend some time examining the logical and political-economic character of programmability and
its historical imbrication with subjectivity, before turning to the narrative
formulations that the programmable (de)subject produces.

Programmability

Concepts of biological, psychological, and social programmability proliferate across the emergence of the control episteme.[5] Like the wider system of
metaphors within which they function, these concepts are grounded in a
specific set of historical shifts centered on the blurring of material and idealized forms of computation. Chun observes that the term *programmability*—
which, unlike the word *program*, came into use only after the development
of the electronic digital computer—"marks the difference between digital

and analog machines."[6] This is the case not because analog machines cannot be programmed, but because where programming an analog machine is descriptive (being concerned with setting up starting conditions, connecting modules, and setting values on a continuous scale), programming a digital machine is prescriptive (being concerned with breaking down mathematical operations into arithmetic steps and thus requiring a machine able to follow a series of coded instructions "precisely and automatically").[7] For something to be considered programmable, then, there must first be a digital formatting of that thing. Chun suggests that the discretization required for digital computing (and thus programmability) equates to a *disciplining* of hardware.[8] As argued throughout this book, such processes of disciplining always include loss, whether they are physically implemented through hardware or conceptually implemented on the range of biological, psychological, and social phenomena that are today viewed through idealized computational metaphors. Chun explains this loss in terms of the digital advancement of counting over measuring that is exemplified in the conversion of a 0.7-volt signal into a 0 and a 5-volt signal into a 1 for the purposes of processing Boolean algebra through switching circuits.

Extending a technical concept to the level of the universal metaphor, the idea of the programmable subject can be seen to undergird many of the signal socioeconomic phenomena premised on the control episteme, including the principles of flexible labor practices, the irresolvable dialectic of deskilling and reskilling, and the vision of social interaction as reducible to communicative acts, as well as the emergent forms of violence that these principles affect. It is telling that Foucault, without any explicit recourse to the notion of the computer as metaphor, describes neoliberal economism as "the strategic programming of individuals' activity."[9] Castells emphasizes this conceptualization of the prospective laborer as programmable when he distinguishes between the principal modes of work in the network society:

Self-programmable labour is equipped with the ability to retrain itself, and adapt to new tasks, new processes and new sources of information, as technology, demand, and management speed up their rate of change. Generic labour, by contrast, is exchangeable and disposable, and co-exists in the same circuits with machines and with unskilled labour from around the world. Beyond the realm of employable labour, legions of discarded, devalued people from the growing planet of the irrelevant, from where perverse connections are made, by fringe capitalist business, through to the booming, global criminal economy.[10]

Castells's three categories—"self-programmable," "generic," and "irrelevant"—broadly map onto Neferti Tadiar's concepts of creative, disposable, and exhausted life, although, like the theorists Tadiar engages, Castells's account lacks a sufficiently heterogeneous concept of life: both telesales workers in the United Kingdom and maquiladora workers in Mexico would fall under the category of generic labor despite the monumental differences in their concrete situations, if not in the underlying logic that produces these situations. Furthermore, Castells fails to account for the way in which all three of his categories are in some way shaped by the concept of programmability: the "self-programmable" worker obviously describes the optimal post-Fordist subject who is able to consciously uphold the ideal of self-steering, work flexibly, and find "creative" solutions to the emergent problems that are characteristic of the network society's dynamic, always-on economy; the very conditions of exchangeability and disposability that define the generic worker are based on the principle that such actors can immediately slot into (and thus be switched out of and replaced) a particular production system or chain—a conceptualization that presupposes a fundamentally programmed and programmable actor, if not one who is able to change their own "programming" as required; and, finally, the "irrelevant" person from beyond the "realm of employable labor" is essentially defined as such by the impossibility of their conceptualization under either of the logics of programmability described earlier, so that their exclusion is characterized as a disconnection that nonetheless does not foreclose on a basic condition of communicability because it leaves open the formal possibility of "perverse connections." The positing of this last fate in the language of communication, despite its supposed exclusion from privileged communicative networks and the programmatic competencies they imply, demonstrates the pervasiveness of control principles within the circuits and shadows of global capital. The ideological foundation of programmability is not limited to the self-programming of the "creative," then, but also undergirds a general conceptualization of the human that facilitates and objectifies a range of social relations.

The figure of the human as programmable object stands as an emblem of control that can today be observed in a range of practices beyond the specific configuration of types of labor. Its diffusion beyond narrow technical and managerial communities is, for example, exemplified by the contemporary belief in and uptake of the pseudoscience Neuro-Linguistic Programming (NLP). The subject of untold publications, seminars, and websites, NLP is presented as a system for success and self-improvement across a range of areas, including business, sales, sports, and the staggeringly misogynistic world of the "seduction community."[11] Developed by Richard Bandler and John Grinder in Palo Alto in the 1970s as a method for psychiatric treatment, NLP broadly consists of techniques for reprogramming both one's own psychological profile (for greater success at whatever one's chosen occupation or pursuit may be) and, in later variants of the system, others' profiles (in order to exert greater influence over them).[12] The discipline is essentially premised on a symbolic, structuralist-linguistic (which is to say, cybernetic) vision of mind: Grinder trained as a linguist, specializing in syntax, and with Bandler he describes the relationship between distinct individual acts (which can be of infinite number) and overall patterns of behavior (which, in their account, are finite, structured, and rule bound) as isomorphic with the relationship between words and sentences, so that syntactical principles can be used to formally model other behaviors.[13] Based on the primacy of this rule-bound, syntactical layer in mediating the subject's relationship with the world, NLP posits optimal patterns of behavior as "programs" that can be either internalized to improve one's own cognitive functions or implemented upon others through verbal and nonverbal "cues" or inputs.[14]

NLP's methodological foundation—the idea that linguistic structure, as an example of complex human behavior, can be scaled to all manner of other behaviors and systems—is remarkable for its similarity to Lévi-Strauss's argument about the applicability of cybernetic modeling to social phenomena in "Language and the Analysis of Social Laws." This similarity, however, seems less remarkable when one notes that the introduction to the book in which this methodological claim first appears is written by Gregory Bateson. Indeed, Bandler and Grinder cite both Bateson and Milton Erickson, core members of the Macy circle since the 1942 cerebral inhibition meeting, as personal acquaintances and intellectual and methodological influences.[15] Bateson went so far as to write an introduction to Bandler and Grinder's book *The Structure of Magic I*, in which he directly posits "'Games Theory [sic],'" "binary information," "pattern and redundancy," and the idea of "homeostasis and self-correction in cybernetics" as precursors to the method Bandler and Grinder propose.[16]

This is not to say that Bateson could have foreseen the progression of NLP into the idealized worldview of the sales executive or the systemic misogyny of so-called pickup artists, or that Bandler and Grinder intended their method (which is dubious enough when confined to psychiatry) to be applied to sales techniques or marketing strategy. Nor is it to argue that the ideological programmability that underpins the post-Fordist labor relations is necessarily the same as that invested in NLP, as if programmability were some actually achievable aim as opposed to an idealized conceptual formation. Either claim would serve to impose a teleological line onto what must be understood as a diffuse set of concepts and ideas that can nonetheless be seen to ground a particular intensification of capital's cybernetic tendencies. Rather, these connections underscore the way in which general concepts of programmability extended to the human body and mind persist across multiple registers—social, economic, and cultural—in the age of control. Furthermore, these connections emphasize the flexibility and range of the human–computer metaphor as well as the drive toward essentialism and exploitation that this metaphor tends to produce when it is extended to socioeconomic phenomena. The concept of "woman" that underpins the NLP methods used by pickup artists, for example, makes Lévi-Strauss's insistence that cybernetic approaches to social phenomena would never lead individual social actors to lose their "character of value, to become reduced to pure signs,"[17] appear either deeply cynical or wildly optimistic. Given the conceptual and cultural differences that exist between the idea of the programmable human and the dominant notions of the self that emerged through psychoanalysis around the turn of the twentieth century, the discursive appearance of programmability can be taken as an index of the control episteme's penetration into the realms of subjectivity and social relations. Following this principle, the upcoming sections in this chapter evaluate the historical emergence of programmability by locating notable transformations in the interlinked concepts of subject and narrative.

Mediated Subjects

Differing concepts of the subject are instructive for thinking through the ways in which technical principles, media formats, and vague metaphors function together with concrete social relations to constitute historical subjects. The basic principles of psychoanalysis and cybernetics, or the depth and surface models that they respectively represent, would seem to place them in diametric opposition as far as approaches to psychology go. This impression is only fortified by observations from theorists such as Deutsch that the emergence of cybernetics represented a historical movement away from "drives" toward a new focus on "steering." These successive theories of the subject, both of which can be neatly associated with contemporary forms of technical media (psychoanalysis with phonographs, cameras, and microscopes, as well as with mystic writing pads; cybernetics with computers and distributed networks) also appear to map neatly onto the disciplinary and control societies that Deleuze places in succession. To adhere to this clean opposition, however, is to ignore both the longer genealogy of control that is partially set out in this book and the work that cultural forms grounded in psychoanalysis and the depth model of the subject continue to perform alongside cybernetic modes of socioeconomic visioning. In the same way that film does not disappear after digital media but continues to play a major role within the cultural landscape of control-era capital, concepts such as desire, drive, and the unconscious continue to perform vital roles within control's cultural layer. Because the idea of universal programmability built on an epistemic digitization (or disciplining) of the subject is a precondition for new forms of subsumption, one should pay close attention to the ways in which this discretization is imposed on and mutates older, nominally "analog" concepts. As argued throughout this book, the mutation of preexisting concepts and formations grants some of the most telling insights into the dominant cultural logic of control societies.

The structure of a given psychoanalytic theory is always contingent on the media technologies extant at the time of its formulation. This is the central claim advanced in Kittler's book *Gramophone, Film, Typewriter* and developed through a specific focus on cybernetics in his essay "The World of the Symbolic, a World of the Machine." Lacan, according to Kittler's argument, is able to unify the three media formats—analog sound recording (the real), film (the imaginary), and standardized writing (the symbolic)—only because the universal computing machine that would go on to erase the distinction between these media had already been invented. By contrast, Freudian psychoanalysis is limited, Kittler suggests, by the technologies available; the Freudian unconscious owes its structure to the fact that Freud could only draw on telephones, radio, phonographs, cameras, and mystic writing pads in order to conceptualize it. After Lacan (or under "high-tech conditions"), Kittler explains, psychoanalysis "no longer constructs psychic apparatuses ... out of storage and transmission media, but rather incorporates the entire technical triad of storage, transmission, and computation." This formulation allows Kittler to move toward a perhaps inevitable claim about the end of the category of the human: the crucial insight of Lacan's reading of Edgar Allen Poe's story "The Purloined Letter," Kittler writes, is that the subject, who is bound to filter the real through the "requirements and exemptions, that is, [the] laws" of the symbolic order, does not think but instead is "programmed."[18]

That these programmed subjects are not a "scandal" in academic and popular discourse, Kittler suggests, is only due to the fact that Lacanian psychoanalysis is "quite deliberately not a natural science," a claim that is justified by the observation that it eschews the continuous linear time of the "clocks which established constants of energy in Mayer and in Freud" in favor of "information machines such as dice, gates, and digital calculators."[19] There is, then, a clear resonance with the historical reformulation of labor in Kittler's claim. As is so often the case, Kittler's account reveals some of the ways in which scientific and technical concepts function as epis-temic things or vapory metaphors that shape historical conceptualizations of social and biological phenomena. Hints at the historical construction of distinctions between natural and unnatural aside, however, his analy-sis stops short of addressing the ever-expanding processes of subsumption under which a concept of the social actor as programmable information processor opens any number of new possibilities for valorization. If one extends Kittler's argument in such a direction (against his stated tendency, to be sure), the cybernetic reformulation of the unconscious attests to the epistemic function of cybernetic logic that, in step with the emergence of new forms of labor, is registered in diverse cultural formations, from bio-logical and artificial materials to socioeconomic principles to concepts of the self.[20]

The programmability that Kittler locates in the Lacanian unconscious takes place within a psychic apparatus that includes real (analog) and imag-inary (mixed analog and digital) registers as well as a symbolic register, in contrast to the Freudian apparatus that (for Kittler) could conceptualize only equivalents to the real and the imaginary. The fact that Kittler brings up (in the introduction to *Gramophone, Film, Typewriter*) the mapping of the Lacanian triad onto the distinct types of technical media alongside the observation that the digital computer removes the distinction between these media suggests that, after cybernetics, the computer and the uncon-scious become thinkable as doubles of each other—a phenomenon that cannot be separated from the epistemic grounding that control-era capi-tal draws from a cunningly reoriented version of cybernetics. Under the cultural logic of control, then, cybernetic principles do not replace psy-choanalysis but repurpose it. Control introduces a programmable object in place of the subject through a process of disciplining, and channels the metaphorical energies associated with the subject into (equally meta-phorical) discrete and thus valorizable quantities. The digitization of desire through the flexible but synchronic system of projected target demograph-ics discussed in chapter 3 stands out as one example of this disciplining process. This capture and valorization of desire within mediating circuits is the subject of Jodi Dean's book *Blog Theory* as well as of Galloway's claim that the dominant function of play under post-Fordism is constituted from a synthesis of romanticism and cybernetics.[21] The argument presented here seeks to bolster such theoretical accounts with a historical grounding cen-tered on the processes through which concepts like desire and the romantic notion of the individual, sovereign subject become thinkable as the energy sources of digital communication systems.

The Unconscious Is Structured Like a Language of New Media

There have been a number of studies of the intersections of cybernetics and psychoanalysis, and rather than reproducing their arguments, the following pages pursue only a single, specific line: the historical process whereby the components of the unconscious developed in Freudian psychoanalysis become thinkable as programs, or discrete units driving physiological or behavioral changes within psychic mechanisms.[22]

It is certainly the case that psychoanalysis occupied an antagonistic role in the early debates around cybernetic psychology. In surveying the intellectual climate from which cybernetics emerged, Deutsch comments that "the depth psychology of Freud and his followers" as well as "the school of *Gestalt* psychology led by Kurt Koffka and Wolfgang Köhler" served to "balance" (that is to say, stand in emphatic distinction to) the materialist conceptualizations of psychology that served as essential precursors to the cybernetic moment.[23] The principal contributors to the cybernetics debates of the 1940s, save for Lawrence Kubie, perceived the principal tenets of Freudian psychoanalysis as mystifications intended to preserve the humanist ideal of the subject against the insights of the most recent fields of scientific knowledge. At the Macy conferences, psychoanalysis was at times used as a measuring post for everything cybernetic psychology was not to be, as Dupuy demonstrates:

The will? All its manifestations could apparently be simulated, and therefore duplicated, by a simple negative feedback mechanism. Consciousness? The "Cybernetics Group" had examined the Freudian unconscious, whose existence was defended by one of its members, Lawrence Kubie, and found it chimerical. If Kubie often found himself the butt of his colleagues' jokes, it was not because he was thought to be an enemy of human dignity. It was rather because the postulation of a hidden entity, located in the substructure of a purportedly conscious subject, manifesting itself only through symptoms while yet being endowed with the essential attributes of the subject (intentionality, desires, beliefs, presence to oneself, and so on), seemed to the cyberneticians nothing more than a poor conjuring trick aimed at keeping the structure of subjectivity intact.[24]

This clash between cybernetics and psychoanalysis at the Macy conferences is recounted in detail by Steve Joshua Heims. In Heims's account, major critiques leveled by McCulloch and Pitts were primarily motivated by the unacceptable levels of inconsistency and ambiguity they saw in the psychoanalytic method, while Bateson was troubled by a quite different problem—namely, the emphasis on consciousness. McCulloch appears to have been most vociferous in his opposition, presenting a paper at the Chicago

Literary Club on February 13, 1952, in which he condemned practitioners
of psychoanalysis as scammers motivated purely by greed and bound to
the method only because it presented opportunities for extended and thus
lucrative treatments. Both Heims and N. Katherine Hayles report that Kubie
responded by framing McCulloch's tirade in psychoanalytic terms.[25]

Hayles presents the McCulloch–Kubie conflict as a relatively straightforward clash between the mechanistic view of cybernetics and the reflexive
character of Freudian psychoanalysis, which for McCulloch represented an
affront to the ethical basis as well as the objectivity of scientific method.[26]
Framing the conflict in such a way, however, risks obscuring the complex
interactions between the two disciplines that can be observed taking place
in the 1940s and 1950s. A more nuanced picture emerges if one examines
a trajectory that begins with Kubie's earlier contributions to neurophysiology and progresses into Lacan's use of cybernetic concepts in the 1950s.
In 1930, before converting to psychoanalysis and becoming affiliated with
what Hayles calls the "hard-line Freudianism" of the New York Psychoanalytic Institute, Kubie published a paper on reverberating loops that would
prove influential on McCulloch and Pitts's conceptualization of the neural
network. While under fire from the majority of the other participants at the
Macy conferences, he presented work combining psychoanalytic insights
with symbolic logic.[27] Wiener, although having "no essential objection to
psychoanalysis," argued that it should be terminologically reformulated
in terms of concepts of system, information, and feedback—a challenge
that Kubie took up in work that prefigures Lacan's adoption of cybernetic
concepts.[28] In short, if one views these discussions of cybernetics and psychoanalysis as a zone of exchange rather than as a clearly defined conflict
between irresolvable positions, a picture of conceptual diffusion emerges
that is instructive in tracing a genealogy of the reformulation of the subject
under the control episteme.

In October 1947, Kubie published "The Fallacious Use of Quantitative
Concepts in Dynamic Psychology," a paper that exemplifies the epistemic,
symbolic disciplining of energetic metaphors into digital ones within
the cultural framing of psychoanalysis. Kubie's essay explicitly connects
the economic phase of Freudian psychoanalysis to energetic metaphors,
bemoaning the use of the latter for encouraging the use of "descriptive
shortcuts" that are antithetical to what Kubie regards as the correct understanding of psychic processes.[29] Kubie opens the essay with an extensive
passage from Freud's *Introductory Lectures on Psychoanalysis* that makes the
role of these energetic concepts quite clear:

You will doubtless have noticed that in these last remarks I have introduced a new factor into the concatenation of the aetiological chain—namely, the *quantity*, the magnitude of the quantities concerned; we must always take this factor into account as well. A purely qualitative analysis of the aetiological conditions does not suffice. Or, to put it in another way, a purely *dynamic* conception of these processes is insufficient; the *economic* aspect is also required. We have to realize (*a*) that conflict between the two forces in opposition does not break out until a certain intensity in the degree of investment is reached, even though the substantive conditions have long been in existence. In the same way, (*b*) the pathogenic significance of the constitutional factor is determined by the preponderance of one of the component-instincts in *excess* over another in the disposition; it is even possible to conceive disposition as qualitatively the same in all men and only differentiated by this quantitative factor. No less important is this quantitative factor for the capacity to withstand neurotic illness; it depends (*c*) on *amount* of undischarged libido that a person can hold freely suspended, and (*d*) upon *how large* a portion of it he can deflect from the sexual to the nonsexual goal in sublimation. The final aim of mental activity, which can be quantitatively described as a striving towards pleasure and avoidance of pain, is represented economically (*e*) in the task of measuring the distribution of the quantities of excitation (stimulus-masses) present in the mental apparatus, and in preventing the accumulation of those which give rise to pain.[30]

If we put aside for a moment Freud's association of the quantitative view of mental activity with economics, which could be productively mapped onto this book's periodization project in which informational metaphors for social behaviors replace earlier energetic ones, we can see that this description of psychic phenomena is particularly telling for the way in which it mirrors Chun's account of analog, nonprogrammable machines as functioning through measuring—"*how large*"—rather than through counting. In proposing a rehabilitation of psychoanalytic principles for the age of information and feedback, Kubie's essay advocates not only a shift from measuring to counting, but also a move from the comparison of fixed components to the consideration of the psychic apparatus as a dynamic system in which relations between components are not determined by the superior size or strength of one over the other. Here it is already possible to identify the contours of a concept of the psyche preemptively attuned to the age of the tech startup.

Crucially, Kubie does not suggest that "every use of quantitative variables to explain psychological phenomena is necessarily incorrect," although he does argue that quantitative explanations based on relative measurement represent a "seductive fallacy to which all psychological theorizing is prone." When in doubt, he writes, "one can always say that some component of human psychology is bigger or smaller, stronger or weaker, more intense or less intense, more or less highly charged with 'energy,' or with degraded energy, and by these words delude ourselves into believing that we have explained a phenomenon which we have merely described in metaphors."[31] In Kubie's intervention, the influence of this quantitative model is such that it is retained in the passage from the economic to the structural phase of Freudian psychoanalysis, so that "writers [by the 1940s] still do not hesitate to talk of strong and weak libidos, strong and weak ids, strong and weak superegos, strong and weak egos."[32] The alternative method Kubie proposes, following McCulloch and Pitts's "Logical Calculus," is a symbolic, networked approach wherein psychological phenomena are viewed as "the result of the interplay of many conflicting intrapsychic forces" so that "any rearrangement of these forces can alter the pattern of the psychological phenomena and can release new forms of overt behavior, without any increases or decreases of hypothetical charges of energy."[33] This formulation does not reject the concept of the unconscious, as McCulloch and others did. Indeed, Kubie's participation in the Macy conferences appears to be marked by persistent claims for the retention of the unconscious within cybernetic frameworks, to the chagrin of many if not all of the other participants.[34] Kubie's 1947 paper instead presents a reformulation of the unconscious so that it is governed by a logic resembling that of switching circuits rather than a logic of comparing or weighing on the basis of analogical concepts such as size or strength. "Energy" within the unconscious here functions in a manner comparable to the McCulloch–Pitts neuron, so that measurable quantities become irrelevant and the simple presence or absence of an impulse governs the release of "new forms of overt behavior." This reformulation is indicative of a version of psychoanalysis in which there is a relationship between "symbolic functions in language formation and in neurosis."[35] In this sense, Kubie's cybernetic psychoanalysis anticipates both the Lévi-Straussian and the Lacanian models of the subject, the latter of which, Dupuy suggests, drew its cybernetic endowment in part from an encounter with Kubie's 1930s work on "reverberating circuits" (or positive-feedback loops) as the cause of neuroses.[36]

Lacan's commentary on cybernetics in his 1954–1955 seminar has occupied a number of theorists and historians. Dupuy's work on the subject has already been discussed and will not be reiterated here. Extensive discussion of this seminar is given in John Johnston's book *The Allure of Machinic Life* and Lydia H. Liu's essay "The Cybernetic Unconscious," the latter of which seeks to locate the source of Lacan's claim that some unnamed cyberneticians were interested in Poe's "The Purloined Letter."[37] For the inquiry pursued here, the significant moments in Lacan's deployment of cybernetic principles occur when he turns to the ego and the formation of the subject in relation to the symbolic order, an analysis that posits the latter in terms that are conspicuously amenable to the configuration of the actor as information processor that defines the control episteme.

As Kittler suggests, Lacan's 1954–1955 seminar attests to the media-historical contingency of psychology in general, and in particular to the emergence of a cybernetically inflected psychoanalysis that does not do away with consciousness, the unconscious, drives, or desire but instead reconfigures them under the figure of a symbolic grid that functions like a governor or steersman to convert the continuous or "analog" real into the discrete, symbolic register of digital programming. This reconfiguration is emphasized when Lacan equates the ego (or the function of the imaginary) to an "electronic tube": "Anyone who's played around with a radio is acquainted with a triode valve—when the cathode heats up, the little electrons bombard the anode. If there is something in the way, the electric current does or doesn't pass, depending on whether it is positive or negative. One can modulate the passage of the current at will, or more simply make it an all or nothing system.... Well, that's what resistance, the imaginary function of the ego, is—it is up to it whether the passage or non-passage of whatever there is to transmit as such in the action of analysis occurs."[38] Beyond the conceptual installation of electronic components within the body that marks an affinity between Lacan's electronic tube metaphor and representations ranging from Hollerith's tabulating machine to Becker's decision unit, this passage is notable for the way in which it posits the ego as implementing an all-or-nothing system upon a current that could alternatively be modulated (an analog process). There is thus, for Lacan, "not the slightest sense of the relation of the ego to the discourse of the unconscious"— the "concrete discourse" in which the ego's switching component merely "bathes and plays its function of obstacle, of interposition, of filter."[39] In this cybernetic psychoanalysis, the energetic concepts that Kittler finds in Freudian psychoanalysis are reconceived as electronic ones—a process that is isomorphic with the disciplining of measurement into counting that Chun discerns as a precondition for programmability. That this digitization of the subject was already being proposed at the Macy conferences and in Kubie's psychoanalytic writings of the 1940s foregrounds the way in which the concept of the unconscious becomes bound up in the complex of historical processes imbricating cybernetic logic; concepts of the human as agent, social actor, and worker; and the process of real subsumption.

In elaborating the role of the symbolic register within his system, Lacan posits an explicitly game-theoretical vision of the world under which all possibilities can be formalized down to binary algebra. "The very notion of cause, when viewed as being capable of bringing with it a mediation between the chain of symbols and the real," Lacan writes, "is established on the basis of an original wager—will it be this or not?"[40] This world picture, however, is not objective and total but filtered. Once again, the exclusion that is central to cybernetic modeling comes into view. The symbolic grid limits the subject's experience of the real to those elements that can be parsed through binary algebra, so that "[a]nything from the real can always come out," but "once the symbolic chain is constituted, as soon as you introduce a certain significant unity, in the form of unities of succession, what comes out can no longer be just anything."[41]

In the light of both Kubie's configuration of Freudian psychoanalysis as a network and the centrality of the concept of programmability within the socioeconomic logic of control, Lacan's evaluation of the concept of the subject through game-theoretical scenarios such as the game of odd or even (which, in the cybernetic view of the world, accounts for all possibilities once they have been through the requisite number of steps of formalization) is particularly instructive. In it one finds the subject reformulated as a black box in a communication system and as a source of statistical possibilities that is analogous to Wiener and Shannon's theories of forecasting in which the singularity of an actual event is "impossible" except as "a transmission of alternatives." Lacan affirms as much when he writes that, "by itself, the play of the symbol represents and organises, independently of the peculiarities of its human support, this something which is called a subject. The human subject doesn't foment this game, he takes his place in it, and plays the role of the little pluses and minuses in it. He is himself an element in this chain which, as soon as it is unwound, organises itself in accordance with laws. *Hence the subject is always on several levels, caught up in crisscrossing networks.*"[42] This vision of the subject as "an element in a chain," playing the role of "pluses and minuses" and "caught up in crisscrossing networks," resembles nothing less than the view of the social actor within Hayekian economics or of the human as node within cognitive or communicative capitalism. It is also clearly prefigured in "the interplay of many conflicting intrapsychic forces," any rearrangement of which can "release new forms of overt behavior, without any increases or decreases of hypothetical charges of energy," that Kubie envisaged in his systemic reformulation of Freud.

This concept of the subject as black box or node is conspicuously germane to the notion of programmability that is a precondition for post-Fordist labor. With the real filtered into discrete steps or on/off states before the subject can apprehend it, the realm of possibility becomes governed by the prefatory limitations that facilitate cybernetic techniques of statistical forecasting. Overt behavior becomes nothing more than an outcome cued by "a throw of the dice in the real."[43] Following this process, the properly unmeasurable fields of possible behaviors and affects are reduced to a finite series of outcomes that can be cued or programmed by symbolic inputs: "Straightforward encoding," Kittler writes, "transfers unlimited chance (the real) into a syntax with requirements and exemptions, that is, with laws."[44] That this programmable "subject" arises within the expanding boundaries of psychoanalytic theory—a set of practices that, although undoubtedly heterogeneous, is rooted in the energetic industrial practices and analog optical and visual media of the turn of the twentieth century and that ten years earlier had appeared conceptually antithetical to cybernetics—stands as a clear marker of the diffusion and penetration of digital concepts that constitute the control episteme undergirding the most recent expansion in capitalist modes of production.

This is not to say that Kubie's and Lacan's psychoanalytic theories actively participate in the expansion of capital in the age of control societies. Rather, these theories are evoked here to emphasize the ways in which the emergence of control principles, grounded in the reformulations of the subject and society examined in the preceding chapters, is registered in the diffusion of its fundamental principles across a range of cultural objects, including critical and clinical theories such as psychoanalysis. One might develop this analysis in a number of directions—for example, by turning to the continued conspiracy-theoretical interest in midcentury mind-control experiments such as Project MKUltra. These semimythical projects, some of which are the focus of Jon Ronson's book *The Men Who Stare at Goats* and the subsequent film of the same name (Grant Heslov, 2009),[45] as well as untold conspiracy sites, YouTube videos, and forum discussions, are far less interesting for their often sketchy revelations about whether human behavior is actually programmable than they are for revealing the historical emergence of a desire for and an interest in such programmability at the same time that capitalism expands and diffuses through the forms of real subsumption discussed throughout this book. In keeping with the principle that guides the second part of this book, that of tracking the emergence and imposition of control through the marks it has left on cultural forms, the next section examines the way in which the concept of the programmable subject is figured within narrative production—that area of aesthetics that, from Aristotle's *Poetics* to the classical Hollywood cinema, appears inextricably connected to depth models of subjectivity.

Narratives of Programmability

Given the fact that Kittler begins his analysis of Lacan and cybernetics with a brief consideration of Aristotle's *Poetics*, it seems appropriate to return now to that text in order to start addressing the specific phenomena that accompany the narrativizaton of the social actor as programmable object. In a neat (if anachronistic) synthesis of narrative form and precybernetic concepts of the subject, Kittler extracts from Aristotle the premise that "the beautiful is defined as that which the eye can easily embrace in its entirety and which can be surveyed as a whole."[46] Using this formula as a guide, Kittler is able to address the form of *Oedipus Rex* through the defining categories of modernist technical media—optics and temporality. "The tragedy of King Oedipus may well arouse pity and fear," Kittler writes, but according to his technical reading of the *Poetics* it is only beautiful because it "fulfills the temporalized optical requirement of having a beginning, a middle, and an end."[47] For Kittler, the Aristotelian tragedy, although a temporal art and thus impossible to perceive in a single moment, can nonetheless be understood as a whole at any point in its unfolding due to the unity and causality of its structure, under which all elements of a given narrative can be gleaned from all others. It is predicated upon the principle that "perception of its form is not resisted by boundlessness."[48] This mapping of the temporal onto the optical, which is also found in the visualizing procedures of cybernetics and system dynamics, marks the progression from subject-driven to object-driven models of aesthetic production that characterize control's cultural penetration.

An image reproduced across several texts on so-called new media renders the elements of Aristotelian narrative in spatial form while making some modifications in order to render these elements applicable to interactive forms.[49] This image stands as an emblem of the difficulty of applying systemic modes of analysis to aesthetic forms that are themselves predicated on nested technical and conceptual systems. Formal models such as these cannot address historical contingencies such as the possibility that twentieth- and twenty-first-century narratives adhering to the classical Aristotelian structure, be they produced under the guidance of the *Poetics* or Robert McKee's *Story*, are optimized around a specific, historically dominant concept of the subject that is displaced by the computer metaphors of control. If one places this pre-control model of the subject in concert with its contemporary practices and conditions of knowledge (such as psychoanalysis, the thermodynamic concept of energy, Fordist production, and analog media), a guiding question emerges: If popular narrative forms that observe the spatiotemporally constituted "beauty" of beginning, middle, and end are imbricated with dominant concepts of depth, linearity, causality, and finitude, then what would be the contours of a narrative form centered on the programmable objects of the control episteme? This question must be addressed not by jumping straight to hypertext fiction or video games, but by tracking the emergence of noninteractive forms that allegorize the digital or digitized subject in ways that produce torturous, conflicting syntheses with the already-established conventions of those forms.

Kittler's formulation of boundlessness as antithetical to Aristotelian ideals of beauty points toward a contradiction that can serve as a guide to this inquiry into the appearance of digitized subjects in noninteractive narrative forms. A contradiction emerges between the necessary temporal limitations on visual or textual narrative forms, which must somehow move forward and eventually end, and the permutational logic of the control episteme, after which one can theoretically be reappointed to some new function at any moment, as in the perpetual present of flexible hours and zero-hour contracts, reskilling, freelancing, and other forms of precarious labor. In other words, the aesthetic products of a clash between (1) linear narrative

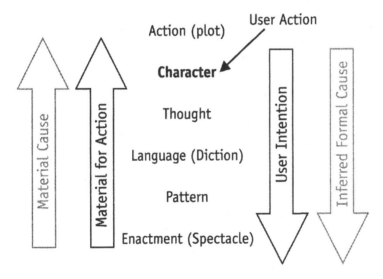

Figure 5.1
"An Aristotelian Framework for Interactive Fiction." *Source*: Michael Mateas, "A Preliminary Poetics for Interactive Drama and Games," in Noah Wardrip-Fruin and Pat Harrigan, ed., *First Person: New Media as Story, Performance, and Game* (Cambridge, MA: MIT Press, 2004), 24.

forms that presuppose a certain stability of characters and readers/viewers alike and (2) the historical appearance of a flexible, programmable object as distinct from earlier concepts of the subject (as either character or viewer) exemplify the cultural reformulation of the subject according to the logical grounding of control-era capital. The tension between the linear and the permutational marks the imposition of control logic into cultural registers.

"Parallax receives demands from all phases of industry, demands for ... unusual personalities. If you qualify, and we think that you can, we're prepared to offer you the most lucrative and rewarding work of your life." This account of real subsumption at work is delivered to Joe Frady, the protagonist of *The Parallax View* (Alan J. Pakula, 1974), by a member of a corporation that recruits and trains assassins for hire. The practices and implications of determining such "unusual identities" as preconditions for employment, appearing as they do in a film that directly works through the relationship between programmability and labor, point toward the interlinked levels of subject formation and narrative form that characterize cultural production in the age of control.

The process for determining the optimal "unusual personalities" for the Parallax Corporation's clients has at least two parts: applicants fill out a paper questionnaire and, if successful, view a montage of words and images in a testing chamber at the corporation's offices. The questionnaire is only partially shown on screen, but the montage is presented in real time and fills the dimensions of the frame for its duration. The questionnaire, or "Personality Inventory," is the initial screening process for the Parallax Corporation and consists of a series of assertions requiring a yes or no answer ("I am a healthy person," "I like romantic stories better than adventure stories," "I am often frightened when I wake up in the middle of the night"). It is a straightforward example of the psychometric test, a filtering of preferences and experiences through binary (logical-symbolic) gates in order to assign a given subject a personality type from a finite, predetermined set of such types.

The montage test is more complicated; it is apparently intended to produce and measure direct physiological responses and thus be impossible to "cheat" by giving predetermined or random (as opposed to "honest" or "authentic") answers, as Frady may well do in the discrete, symbolic register of the written test in order to gain access to Parallax. The montage test requires the subject to sit in a chair with their fingers pressed to a pair of electronic devices while a series of words (*love, mother, father, me, home, country, God, enemy*) are juxtaposed with a series of images that become increasingly jumbled and are cut together with increasing speed as the montage progresses. It is not clear whether the montage is another, more advanced type of psychometric test or a method of programming (so-called brainwashing). The latter would place the film in the tradition of "sleeper agent" texts that emerged from the late 1950s on—a tradition that includes Richard Condon's novel *The Manchurian Candidate* (1959) and its two film adaptations (John Frankenheimer, 1962; Jonathan Demme, 2004), the film *Telefon* (Don Siegel, 1977), and the fourth season of the television series *24* (Fox, 2005)—and that are clearly grounded in the prospect of programmable behavior that emerges from cybernetic approaches to psychology and shadowy projects such as MKUltra. In any case, whether the montage in *The Parallax View* is a test or the installation of a program, it is interesting primarily for the connection it draws between form and narrative function. It densely interweaves material directed at the intersections of the symbolic and the imaginary and in so doing suggests that short circuits between these registers can function to diagnose (and possibly alter) the steering mechanisms (or programs) of the subject, at least in ways that can progress a plot centered on the behavior of such a subject.

Whether the Parallax montage is a test or a mind-control program, its steering function is borne out in the film's narrative structure. Although many requirements of a complete narrative arc—an identifiable antagonist, a fully graspable conflict, and a clear resolution—never come into view, what does become evident across the film's duration is the premise that the corporation deploys both assassins and patsies and that Frady has in fact been recruited as the latter.[50] The Parallax tests identify personalities in a manner that mirrors the sorting processes of real subsumption and post-Fordist businesses. The assassins, flexible, creative, and reusable labor, are presumably individuals who produce certain (unspecified within the film) results in the personality tests, although not necessarily individuals who reveal violent characteristics. The patsies, who may also be identified via the same tests, oscillate between generic and disposable labor: they are required because their "innate" characteristics enable them to be cued toward certain behaviors and to be framed as nonconspiratorial assassins, but they do not require any particular aptitudes (because they do not commit actual assassinations) and are unimportant enough to be used once and then killed to protect the real assassin, the corporation, and their clients. Because the recruitment process is only represented diegetically through Frady's involvement with it, it may even be the case that the assassins are recruited a completely different way and that only the patsies who mask the assassins' activities are recruited through the psychometric tests. Indeed, the Parallax Corporation might locate and recruit only patsies, with the assassins drawn from an entirely different agency. Either way, the film certainly presents a narrative in which more or less disposable and precarious labor is deployed, without any choice on the worker's part, based on the testing and cuing of behavior through questionnaires and semiotically shaped images—a synthesis that preempts the recruitment and working patterns of "creative" post-Fordist labor, the many valences of freelance and zero-hours contracts, and the valorization of looking and clicking that underpins the attention economy of web and visual culture.

It is not clear precisely when in the narrative of *The Parallax View* the recruitment process begins, but it accelerates after Frady enters the apparatus of electronic sensors and screened montage. The final moments of the film—in which a Congressional special committee announces that Frady, now dead at the hands of a Parallax agent, acted alone to commit an assassination that (the viewer knows) was in fact carried out by a Parallax assassin—reveal that his "unusual personality" made him ideal for the role of the patsy, not the assassin. Impulsive and aggressive character traits measured though yes/no responses that can be easily faked and through physiological responses to verbal-visual stimuli that cannot—these data make him an optimal subject both to steer into a particular situation and to portray as a sociopathic loner who carries out a political assassination without support and thus does not implicate any others. Whether this personality was fully diagnosed by the montage-feedback apparatus or the latter simply served as a step in his progressive entrapment/deployment, it becomes clear at this point that Frady is "cued" to move from step to step until he is in the exact time and place, both geographically and psychologically, to be framed for the assassination. To reiterate Kittler's paraphrasing of Lacan, Frady does not think his way through the sequence of events that compose the narrative of *The Parallax View*. Instead, his investigation—a pursuit that Lacan's analysis of "The Purloined Letter" suggests has moved from a humanistic process of rationality to a process of statistical forecasting—follows the discrete steps of a program until it comes to a halt.

The form of the programmability narrative presented in *The Parallax View* is not a unidirectional arc but a series of decisions that can turn in any direction—an observation Jameson preempts when he observes a midcentury turn toward spy thrillers in which plots revolve on the "facile but effective device of the double agent, so that teams of villains can be transformed into heroes at the flip of a switch." This switching-circuit-as-narrative is driven by a logic of permutation, a deployment of individual characters as interchangeable units within a formal structure that goes "a certain way towards declaring at least the intent to construct a narrative which is in some way an *analogon* of and a stand-in for the unimaginable overdetermination of the computer itself."[51] Jameson's claim is compelling and can be extended if one views the computer not only as an exemplary tool of late capitalism but also as a source of metaphors that underpin the configuration of the human actor caught in the networks of this political-economic epoch. Although the function of programmable characters is one of switching, filmic narrative is still bound to linear forward movement, leading to conspicuous narrative artifacts that betray the untimely logic of control. Jameson suggests as much when he points out that the outcome of implementing such narratives in film and literature is "confusion rather than articulation"; the production of an aesthetic based not on the Aristotelian ideals of unity but rather on the informatic conceptualization of the subject leads to a situation where the viewer/reader is "no longer able to remember which side the characters are on, and how they have been revealed to be hooked up with the other ones."[52] In *The Parallax View*, this switching logic of the double agent is abstracted from a conscious to an unconscious register so that the protagonist (the detective-rebel) is steered toward reinforcing the conspiracy while performing actions he believes will unveil it. According to Jameson, "it is precisely the will to revolt and to destroy the conspiracy which allows this last [the conspiracy] to write him into their scenario.... The detective is thus murderer and victim all at once."[53] The concept of the subject in this narrative formulation "plays the role of the little pluses and minuses" within a game of odd or even. And narrative, like the concept of subject formation after the imposition of the symbolic register, becomes an analogon for the "throw of the dice in the real."

The distance between *The Parallax View* and the "sleeper agent" type of story, of which *The Manchurian Candidate* is exemplary, can be located in the way in which the former marks the viewer (as well as one or more characters within the diegesis) as potentially programmable. The real-time, full-screen presentation of the Parallax Corporation's montage sequence attests to this marking of the viewer. An edited sequence of images and words designed to scramble clear associations within a story that, although causal, turns on the oscillation of opposite responses (fascination/compulsion and confusion/boredom) suggests an emerging awareness that the premise of programmability functions not only as a narrative device or object of representation but also as a concept directed at a projected viewer and executed through aesthetic forms such as narratives.

The idea that moving images might be understood as a form of programming is supported by Kittler's theorization of such images as a mixed analog and digital format. On the one hand, Kittler places film in the category of

analog technological media that, unlike writing before them and computers after, is able to capture material that exceeds symbolic representation.
Sybille Krämer has summed up this aspect of Kittler's position as an intervention into the concept of media history:

> In the era of writing, one could only write things down that already existed as ele-
> ments in the symbolic universe—or in other words, the things that are inherent to
> the "nature" of a sign—but after the technological analog media have broken writ-
> ing's monopoly, one can record the extra-symbolic—or that which is beyond the
> symbolic realm. In other words, one can record nature itself. Technological media
> allow one to select, store, and produce precisely the things that could not squeeze
> through the bottleneck of syntactical regimentation in that they are unique, contin-
> gent, and chaotic.[54]

In Kittler's system, however, cinema does not equate perfectly to the real (a
status he affords only to analog sound recording). Because of the illusion of
movement on which it is premised and the centrality of editing to its dominant applications, Kittler places film between the real and symbolic, so that
it appears a double of the imaginary.[55] Since Thomas Edison's acquisition of
the patent for a "so-called escapement disc mechanism," which "ensured
that the individual frames for the film stood beautifully still during the
sixteenth of a second in which they were recorded or observed, while all
further transport between the individual frames fell precisely in the pauses
in between," Kittler writes, film has been "a hybrid medium that combines
analog or continuous single frames with a discontinuous or discrete image
sequence."[56] This is, of course, a realization that can only be made after
the emergence of the electronic digital computer as universal metaphor. If
photography is analog and editing is digital, then the assemblage of both—
film—captures movement in discrete cells and converts continuous motion
into input for the viewing subject. This historically specific understanding
of film equates to the reconceptualization of desire as steering that characterizes the cybernetic reformulation of psychoanalysis, as well as von Neumann's account of the brain as a mixed analog/digital system in his theory
of automata. In the age of control, moving images come to be understood
as an aesthetic analog to the disciplining of hardware that is a necessary
precondition for the latter's programmability.

A continuing movement from content to form in the representation
of programmability is supported by Galloway's concept of disingenuous
informatics. Discussing what he calls the "control allegory," Galloway suggests that films of "epistemological reversal"—such as *The Usual Suspects*
(Brian Singer, 1995), *The Game* (David Fincher, 1997), *Fight Club* (David

Fincher, 1999), and *The Sixth Sense* (M. Night Shyamalan, 1999)—emerge in the 1990s and 2000s in place of the political conspiracy thrillers that Jameson analyzes in *The Geopolitical Aesthetic*. Narratives of epistemological reversal, for Galloway, "aim at doling out data to the audience, but only to show at the last moment how everything was otherwise."[57] In other words, the narrative function of these films is not premised on continuity, nor on the subject-led, mythic character arcs of classical commercial cinema, but on a logic of selection and exchangeability under which false information is provided up front purely to allow for climactic moments of revelation in place of narrative closure or catharsis. Here anagnorisis does not lead to peripeteia, as it does in Aristotle's definition of the optimal tragedy. Instead, the moment of revelation marks the climax and the conclusion of the narrative, and more often than not is directed at the viewer rather than at a character internal to the plot. This appears to be less a break from the formal logic that emerges in *The Parallax View* than an intensification of its programmatic engagement with the viewing subject. Both are premised on a narrative logic in which anything can be exchanged for anything else so that a single frame, line of dialogue, or word (in this case the point of informatic revelation) can function as a switch that transforms the significance and structural function of what came before it. In the case of the epistemological reversal plot, this switching moment becomes the sole ordering principle, so that the film functions not as morality tale or a mirror for the reflection of the viewing subject's desires but as a selection from a network of interchangeable possibilities. This procedure can be seen at its most intense in the formal techniques of contemporary action cinema.

Form as Programming

Following the progression traced in the preceding material, in which the reticle of programmability is increasingly trained on the viewer of late twentieth-century cinematic spectacles, contemporary action cinema can be understood as part of a filmic system that functions to execute particular routines in the viewer, meeting predetermined requirements with the correct inputs to excite them. An executive mode of editing can be contrasted to both the obfuscating function of continuity editing (in which cutting is subject to a strict set of rules designed to obscure its artificiality) and the interpretive mode of montage (in which a comparison between two series of images is invited). But, once identified, this executive principle can also be seen to operate beyond the register of editing techniques. It can be located, for example, in continual camera movement along the x, y, and z

axes—examples of this technique abound in Paul Greengrass's *Bourne* films (2004, 2007) and in Mark Neveldine and Brian Taylor's *Crank* (2006), *Crank 2* (2009), and *Gamer* (2009)—as well as in the use of digital effects, virtual camera techniques, and cameras with very high frame-per-second capture rates to create longer, faster, and more spatially commanding pans and tracking shots. The critical element in this style is measurable action, a perpetual motion of or within the frame that replaces the formerly dominant imperative of continuity. David Bordwell defines the effect of action editing schemes in the works of such directors as distinct from the more general tendency toward "intensified continuity" that he locates across contemporary American commercial films. Bordwell suggests that the effect of much contemporary film style is a "vague busyness, a sense that something really frantic but imprecise is happening," and Steven Shaviro suggests that in "postcontinuity" films "shots are selected and edited together only on the basis of their immediate visceral effect," with "no concern for any kind of pattern extending further in space and time."[58] Similarly, Sean Cubitt notes that such films as *The Rock* (Michael Bay, 1996) and *The Matrix* (Lana Wachowski and Andy Wachowski, 1999) "feature major scenes in which the geography of the diegesis is radically unclear."[59] Given this elevation of action over continuity, this executive mode of commercial film might be described as "feedback cinema," following Bateson's use of the phrase "[the] difference that makes a difference" to group "ideas," "bits," and "unit[s] of information" within communication-theoretical notions of organisms and social bodies.[60] In the feedback model of cinematic form, technique is no longer subordinate to demands of continuity and narrative but is instead deployed to introduce regular perceptible difference for an ideal viewer who has been historically conceptualized only to register such differences.

This executive formal mode can be understood as an intensification of the process that Jean-François Lyotard describes as "the oppression of orders." In a short essay titled "Acinema," Lyotard suggests that continuity editing "protect[s] the order of the whole" within the commodity-aping logic of classical commercial cinema by defining and eliminating mistakes before the film object is finalized and released for public consumption, thus "banning" the intensity a perceptible mistake might carry in the context of an otherwise continuous and perfected form. Lyotard connects this oppressive protection of order to the flow of capital through the metaphor of the struck match: the causal sequence of the match struck and thus expended in order to "light the gas that heats the water for the coffee which keeps you alert on your way to work," expressed as the cyclical formula merchandise/match → merchandise/labor power → money/wages → merchandise/

match, is contrasted to the "sterile differences leading nowhere" of the child who strikes a match just for the fun of it. Lyotard takes great pains to remind his readers that the replication of this "pyrotechnic" causal system in continuity editing is not propaganda (films designed to "lull the public consciousness") but rather propagation ("the effort to eliminate aberrant movements").[61] In marked contrast to the commercial cinema that Lyotard addresses, cinema in the age of control effects a situation whereby the crucial distinction for the sublimation of capital flows is no longer between "ordered" and "aberrant" movement but between movement and non-movement or between measurable behavior and everything else. Watching films directed by Bay, Anderson, Neveldine and Taylor, or McG (Joseph McGinty Nichol), one imagines that even the "wretched" ideology-focused critic that Lyotard decries would find it difficult to argue that these films' aim is to "lull." After the historical reformulation of the subject as information processor, the objective of commercial media is no longer to opiate but instead to stimulate in a way that doubles the relationship between input, black box, and output in an idealized electronic communications network.

The stimulatory logic of execution foregrounds the principle that contemporary cinema should not be analyzed solely in terms of a Virilian narrative of ever-increasing speed, at least as measured on linear scales such as numbers of shots or average shot length. Indeed, the most recent stylistic orthodoxy in the Hollywood blockbuster (seen, for example, in *The Avengers* [Joss Whedon, 2012] and *Transformers: Dark of the Moon* [Michael Bay, 2011]) appears to signal a move away from fast cutting between often unmatched shots and toward the use of artificial long takes composited from multiple shots, each of which contains rigorously maximized levels of movement—of the frame itself, of characters, of objects, of particles of debris such as sparks and broken glass, and of fire and water. The rigor and intensity of (1) developing marketing systems and modeling an audience for these cultural objects based on past activities and preferences; (2) optimizing the final filmic object for this projected audience through extensive principal photography, the creation of computer-generated effects, and the compositing of live and computer-generated elements; and (3) the post-production and editing workflows characteristic of high-budget productions all emphasize the purposefulness with which these perpetual-motion spectacles are created in order to meet a particular, historically specific concept of the viewing subject.

If the passage from the late 1950s to the 2000s marks the emergence of the concept of programmability in both subject matter (*The Manchurian Candidate*, the NLP craze) and narrative form (intensifying from *The*

Parallax View to the "epistemological reversal" narrative of films such as *The Sixth Sense*) as well as the emergence of an executive style that seeks to provide maximum informational input for an idealized (and thus programmable) target viewer, the past decade or so has seen the appearance of a number of texts that synthesize all of these phenomena. The recurring figure of the clone or artificial human with an implanted personality and false memories attests to an ongoing interest in subject matter centered on the programmability of social actors, and can be found in all manner of science fiction texts, from *Blade Runner* (Ridley Scott, 1982) to *The Island* (Michael Bay, 2005) and *Moon* (Duncan Jones, 2009). In each of these stories, implanted memories and behavioral patterns are central, emphasizing literal programmability (biological and/or psychological) and thus gesturing toward the historical production of the social actor as intelligible through the principles of informatic control. In terms of the synthesis of subject matter and form, though, the first three *Bourne* films (*The Bourne Identity* [Doug Liman, 2002], *The Bourne Supremacy* [Paul Greengrass, 2004], and *The Bourne Ultimatum* [Paul Greengrass, 2007]) are exemplary in drawing together the distinct elements of the historical arc described in this book at the level of aesthetic formulation.

In *The Bourne Identity*, an amnesiac is pulled out of the Mediterranean and sets out on what will become a three-film journey to recover his identity—he is a highly conditioned "sleeper agent" produced by the CIA "black ops" program Treadstone. Within this generic narrative arc, a multivalent image of the society of control is constituted. Critically, these sleeper agents are not clones or genetically modified humans, but "normal" people who have been invested with specific training and then embedded in society to await "activation." They are both switchable and optimized to the psychological and physical demands of their work. These agents' programmability thus points to a kind of control realism in which the ideological penetration of programmability is played out at the dual levels of subject and system. At the level of the amnesiac protagonist who becomes known as Bourne after assuming one of the multiple identities he possesses documentation for, it is possible to locate a vision of affects, skills, and competencies that are programmed and subject to triggering by symbolic or affective input. This is first apparent in a sequence near the start of *The Bourne Identity* in which the still-disoriented and directionless Bourne fights two Zurich police officers who awaken him and attempt to arrest him for sleeping in a public park and lacking citizenship papers. The switching of registers from passive—"I just want to sleep"—to hyperkinetic, the instantiation of a flurry of movement that appears to surprise Bourne as much as the police officers, presents bodily virtuosity as programmed and executed rather than learned and enacted. This is a vision of the body as a form of the disciplined hardware that Chun identifies as synonymous with the concept of programmability.

This presentation of character as programmable object is doubled at the formal level in the *Bourne* films, extending the control conceptualization of the human from the figure of Bourne himself to the projected viewer theorized earlier in relation to executive style. Across the series of films, it is possible to observe an intensification of this style, particularly in action sequences wherein high levels of handheld camera movement and nonlinear editing produce an impressionistic mode so that the general presence of action, rather than the specificity of particular actions in a causal sequence, is primary. That Bourne emerges so consistently as the victor in these action sequences only foregrounds the tension between the imperatives of character-driven plot and the cybernetic conceptualization of the social actor under which subjectivity and embodied knowledge are less deep-seated, socially negotiated formations than a Shannonian field of probability. The actual event (Bourne wins) is at the same time nothing more than a selection from the field of coded, interchangeable possibilities and the only selection from the field of possibilities that can make it onto film. The ideal viewer presumed by this type of image regime does not synthesize impressions or images but rather, like Wiener's antiaircraft predictor, tracks movements and records outcomes.

Finally, the narrative trajectory produced by this deployment of protagonist as program and of form as execution is that of the Markov chain, in which only states n and $n + 1$, the present and the next state, are operative.[62] In this narrative mode, amnesia, that hallmark of unimaginative writing, becomes a formal necessity because it allows the reflexive or reactive logic of the programmable object to progress unburdened by either conscious memory or unconsciously recorded trauma. There are flashbacks across the *Bourne* films, but they are confined to depicting plot information relating to the hit gone wrong that set the conspiracy in motion. Compare these flashbacks to those found in *RoboCop* (Paul Verhoeven, 1987). Whereas Alex J. Murphy, an amnesiac character who has been programmed in the literal, computer-scientific sense after being killed and resurrected as a cyborg, begins to remember his family, Bourne can only flash back to information that is significant to the immediate narrative situation—his past hits and the circumstances surrounding his memory loss. Memory, the accrued, perpetually forward momentum of action sequences across the series of *Bourne* films suggests, is not only separate from but in fact an impediment to bodily expression, virtuosity, and movement. That the narrative consisting of the perpetual forward motion of a programmable object, Bourne, is intercut with a second arc consisting of scenes of CIA analysts in cramped, computer-filled offices and archives as well as in corridors, cars, homes, and hotel rooms, electronically tracking Bourne's activities in order to reacquire him, only emphasizes the purely informatic, game-theoretical concept of subject and narrative that the films enact through the figure and trajectory of Bourne. Indeed "Blackbriar," revealed in *The Bourne Ultimatum* as the umbrella program under which all secret CIA operations sit and thus the protocol from which all of the series' prior events stem, exists to facilitate real-time decision making up to and including the expedient assassination of human targets, its very existence providing a diegetic geopolitical and infrastructural analog for the digital, networked modes of behavior and/as psychology that the films present.

Between these two interlinked assemblages of spatial diagram and narrative arc one can identify a tentative vision of the dialectical mapping of late capitalism that Jameson called for, now repurposed to account for the imposition of a digital epistemology that governs both idealized concepts of system at the global level and idealized concepts of the social actor at the local level. That the subjective experiences produced at both of these levels—the informational labor of the analysts and the constant virtuosic responsiveness of Bourne—present completely different forms of twenty-four/seven activity, perpetual precarity, and ever-present fear of informational disappearance speaks volumes. The extension of these ideals of programmed and programmable behavior to the viewer who is imagined as subject to the constant stimulation of post-continuity editing, camera movement, and diegetic activity—the proliferation of differences that make a difference—attests to the fractal character of this cultural formulation of epistemic conditions. Whether in psychoanalytic theory or narrative form, the political unconscious of control turns out to be paradoxically centered on the reformulation of the unconscious as an information processor.

Coda: Interactivity as Programming

For the most part the present inquiry has avoided engagement with interactive media, predominantly for the methodological reasons discussed in chapter 3, but there is some value in detouring into the analysis of such forms at this point. Although the principal aim of this book has been to locate representations of the often ill-fitting implementation of a digital ontology within areas of concrete social reality that are not immediately directed through computer interfaces, the concept of the Markov chain as narrative form worked through in the preceding pages describes nothing if not the experience of interacting with digital media forms such as video games. The continuity of this claim with the historical-formal principles detailed throughout this book can be emphasized through a consideration of the respective roles of film editing and user input across distinct media forms.

In the second of his cinema books, Deleuze defines two classes of moving image, each of which is produced through distinctive editing schemes. He characterizes these classes, distinguished by the rational and the irrational cut respectively, in the following manner:

[In classical cinema] the cuts which divide up two series of images are rational, in the sense that they constitute either the final image of the first series or the first image of the second ... when there is a pure optical cut, and likewise when there is false continuity, the optical cut and the false continuity function as simple lacunae, that is, as voids which are still motor, which the linked images must cross. In short, rational cuts always determine commensurable relations between series of images.

[In modern cinema] the image is unlinked and the cut begins to have an importance in itself. The cut, or interstice, between two series of images no longer forms part of either of the two series: it is the equivalent of an irrational cut, which determines the non-commensurable relations between images. It is thus no longer a lacuna that the associated images would be assumed to cross: the images are certainly not abandoned to chance, but *there are only relinkages subject to the cut, rather than cuts subject to the linkage.*[63]

The rational cut predominantly maintains continuity between series of images, whereas the irrational cut draws attention to itself by breaking continuity, subordinating the series of images to the now conspicuous formal principle of editing. Each of these image regimes furnishes a mode of narration: mapped onto Deleuze's distinction between historical regimes of cinema, the rational cut produces the movement image, whereas the irrational cut produces the time image. It should be noted that this is also a distinction, in Deleuze, between classical and avant-garde modes of production—put simply, the movement image belongs to the former and the time image to the latter. These forms are thus consistent with the two types of commercial cinema theorized earlier in this chapter. As Lyotard argues in "Acinema," the movement image of classical, continuity cinema aims to produce a continuum of attention, where no component emerges that would draw a viewer's attention to the artificiality of the experience. And as the passage from *The Parallax View* to the *Bourne* films suggests, the emergence of the concept of the social actor as programmable object in place of the psychological or psychoanalytic "deep" subject can be aesthetically located in the decline of this continuity mode in favor of a cinema of bits, or differences that make a difference.

But what kind of aesthetic form results from a mediatic arrangement whose principal function is to stimulate not just attention but also physical action and measurable input? For this input-output loop structures the demand of digital media, whether in the form of the video game, the Facebook profile, or the type of interactive advertisement that compels the user to perform some trivial task. The advertising images that inform Debord's theorization of spectacular culture implore the viewer to respond, to buy the advertised product, but at some time in the future. Such images function, as Edward Bernays's repurposing of Freudian psychoanalysis shows, in relation to the classic psychoanalytic concept of the subject even as they participate in the remaking of this subject according to the emerging demands of control. The images produced by software interfaces, video games, and websites demand an instant response, the type of motor response so fascinating to the 1940s cyberneticians, which in many cases functions more quickly than the time needed to consciously process and interpret the image.[64] This type of image cannot be described as a movement image (obfuscating cuts and smoothing the temporal axis) or a time image (revealing and fragmenting the temporal axis). Each significant change in the image is directed not only at the construction of a specific mode of perception, but also at the motivation of some form of input—a mouse click, a keystroke, or a button press.

One might here recall Kittler's rejoinder—made in his essay "Computer Graphics: A Semi-technical Introduction"—that although modern graphical computing appears premised on user interfaces that directly reference the lineage of photography, film, and television in their appearance and relations, the fading memory of a computer screen populated by nothing but white dots on an amber or green background should remind us that the "techno-historical roots of computers lie not in television, but in radar." The crucial aspect of the radar image, of course, is not the visual form it takes but the fact that the user "must be able to address the dots, which represent attacking enemy planes, in all dimensions and to shoot them down with the click of a mouse."[65] The image in this setup may present a model of the world or a model with no correlate that the viewer can identify in the object world outside of the simulation, but in any case it is always aimed at motivating user action. It is a program image, an image that at once executes and is executable.

Rather than the structure of Aristotelian drama, it is this ideal of the near-simultaneous executability of text (or image) and viewer that historically undergirds the narrative form of interactive media. This relationship must be seen as an outcome not only of technological possibility but also of the historical emergence of the control vision of the social actor. Wolfgang Ernst emphasizes the tension that this imperative of execution introduces to narrative form when he describes the temporality of interactive media as a von Neumannian superposition of computer time and human user:

Culturally the most common way of information processing is the human–machine communication (and its time-critical escalations such as computer games). The temporal constellation which has replaced the narrative, dramatic aesthetics of (tragial or happy) ending, for computer games and human–computer interaction in general, is the mode of interrupt. Thus, kairotic time replaces chronos. Such interactive events between computer and human unfold in idiosyncratic time (based on the "interrupt" mode of computing) rather dynamically than algorithmically, in contrast to the computational steps that unfold within the computer itself, where instruction–execution histories express an ordering of inner events of an algorithm without any relation to the actual passage of time.[66]

This experience of interrupted temporality, which for Ernst appears in dialectical tension with the form of tragic drama that provides the normative mode of commercial narrative production with its exemplar, results from the meeting of human and computer time, or the clash between the desire for and the conventions of narrative and the halting imperative of the algorithm. Wark makes a similar claim in *Gamer Theory*, writing that under the historical conditions of gamespace, which are isomorphic with the conditions of control, the "play that can never end"—the analog diagram of Sisyphean labor—is replaced with "the game that stops, and starts, and stops, and starts, and stops, and starts—forever."[67] This algorithmic structure accounts for the contradictions exhibited by video games incorporating Aristotelian narrative arcs, but at the global scale it also describes the imposition of computer time onto human sociality, from the physically and mentally unbearable working schedules of maquiladoras and other free-trade-zone production facilities to the perpetual present of the zero-hours contract, to the diffuse time of the twenty-four/seven working week facilitated by portable communications, to the inhuman speeds of high-frequency trading. It describes the imperative of real subsumption, under which identity, thought, and sociality are either captured or determined by grids of clicks and keystrokes.

"They don't do random. There's always an objective, always a target." So states the CIA analyst Nicky Parsons of the programmable objects such as Bourne that Treadstone believes it produces. Beyond this statement's literal description of the conditioned assassin and even its allegorical articulation of the type of executive narrative mode that drives the *Bourne* trilogy and that finds its technically supported apotheosis in the temporal experience of video gaming, what better way to describe the social actor as conceptualized under the control episteme? After all, Kittler, whose technological a priori is in this book deformed into a subjective point of view that is intelligible as a double of that attributable to capital itself, equates the displacement of the subject by the computer with a conflation of targeting and programmability, or self-steering: "bees are projectiles, and humans, cruise missiles," Kittler writes, because "one is given objective data on angles and distances by a dance, the other, a command of free will." What is critical about this claim is that its ballistic conceptualization of sociality rests on historical conditions under which "[e]lectronics ... replaces discourse, and programmability replaces free will."[68] Sociality, from the epistemic position that grounds the control era, can be understood only as targeting under the continued impression of free will—a conceptual frame underscored by the fact that the terms *reticle* (gun sight) and *network* share a root in the Latin *reticulum*, "net."

In the first-person shooter game, such as *Half-Life, Deus Ex, Halo,* or *Call of Duty,* it is conventional for the targeting reticle to be the point through which all interaction with the modeled world is channeled, so that one aims a gun sight at a door to open it and at a person to speak with or heal them. It is only in *Metroid* and its ilk than one shoots a door to open it, a fact that, in the context of the kairotic, switching logic of narrative worked through earlier, calls to mind Lacan's theorization of the door as a flip-flop circuit between the real and the symbolic.[69] Equally, in role-playing games such as *World of Warcraft* one heals or gives gifts to others through the exact same procedure of highlighting and selecting options from a drop-down menu that one would use to attack them.[70] If there is a mode of interpretation through which the grids of digital media can be seen to articulate the social imaginary of control, this is it. Under the control episteme, targeting—the practice from which Wiener first developed the concepts of control and steering and which, for Kittler, equates to the conflation of free will and programmability—becomes the horizon for all possibility.[71] This connection emphasizes the principle that the military origins of cybernetics are not the source of its malign implications. Rather, the self-steering gun sight of Wiener's and Bigelow's ultimately unsuccessful wartime experiments represents but a single, telling realization in electrical components and metaphors of the digital concept of pure relationality that already informed nineteenth-century dreams of political economy (Babbage) and governance (Hollerith). Post–World War II cybernetics, as argued in the first part of this book, simply provided an epistemic foundation for the extension of this dream of capital to the "rich and random multiple realities of concrete everyday experience," be they pleasurable, mundane, or, as is too often the case, agonizing.[72] Living in a world whose conditions of possibility are set by this perspective, one no longer thinks or desires or interacts or does nothing unless these practices can be understood as communication, which is to say as targeted or purposeful behavior. Nor, as a consequence, are human beings truly understood to suffer the physical and mental wounds of exclusion, incarceration, or exploitation. One targets or is targeted, or else there is no "one."

This exclusion of life that neither targets nor is targeted appears as a revolutionary potential in a number of theoretical works, from Deleuze and Guattari's characterization of the *hors-classe* to Hardt and Negri's multitude.[73] This totalizing valorization of the excluded is at least as troubling as it is inspiring, and because of this it is necessary to distinguish between those individuals and groups that are systemically excluded and those that are premised on the intentional evasion of capture. Just as the lens of a digital camera might admit as much light as an equivalent lens on a film camera, with digitization (and thus loss) taking place within the image sensor of the camera itself, many of the forms of life rendered superfluous to capital are first captured and apprehended. The setting up of the threshold between valorizable and surplus life thus constitutes one of the principal forms of violence inherent to control, and the problems in claiming the surplus—a category that is by definition not formed by its members' volition—as a revolutionary force (as Deleuze and Guattari and Hardt and Negri do) are clear enough. The status of these forms of life is comparable to that afforded to noise in cybernetic systems: a theoretically countable part of the field to be modeled, but one that does not fall within any defined (and thus prejudged as useful or necessary) logical type or algorithm.

There is, however, a second type of invisibility inherent to cybernetic methods, one that, although potentially deleterious, foregrounds a limit to control that is based on the impossibility of capturing certain phenomena in the first place. It may be that locating the tensions that form around this limit can point to potential modes of resistance that do not rely on the purely conceptual (and non-volitional) valorization of the excluded and expelled. Deleuze himself writes of finding "new weapons" in order to counter the impositions of emerging capitalist formations, proposing "viral contamination," "piracy," and "noise" as the active and passive forms of this emerging counterpractice in contrast to the "entropy" and "sabotage" that might have posed threats to prior modes of exploitation and injustice.[74] He calls for "vacuoles of noncommunication" that might enable one to "evade control" and for "new ways of thinking" that "aren't explicable in terms of microsurgery."[75] Agamben's book *The Coming Community*, originally published in the same year as Deleuze's "Postscript," presents a conceptualization of "whatever singularities," beings "radically devoid of any representable identity," "appearing as uprisen beings ... neither blessed like the elected, nor hopeless like the damned," and thus "absolutely irrelevant to the state."[76] Tiqqun, in "The Cybernetic Hypothesis," end their extensive theorization of the connections between cybernetics and capital by insisting on fog as the "privileged vector of revolt" and affirming the importance of "desires that exceed the flux insofar as the flux nourishes them without their being trackable therein," desires which "pass *beneath the tracking radar*, and occasionally establish themselves."[77] Galloway and Eugene Thacker claim in *The Exploit* that "future avant-garde practices will be those of nonexistence" in the form of "nonexistent action," "unmeasurable or not-yet-measurable human traits," and "the promotion of measurable data of negligible importance."[78] Chun has written on the need to "exhaust exhaustion, to recover the undecidable potential of our decisions and our information through a practice of constant care" as a response to the situation of crisis as norm that neoliberalism has effected.[79]

These theoretical formulations are but a representative sample of the many comparable accounts of escape that have appeared since the publication of Deleuze's theorization of control societies in the early 1990s. They relate to each other in several ways. Most obviously, they are concerned with states of undecidability or unmeasurability, and most rest on a fundamental identification between computer technologies, cybernetic logic, and late twentieth- and early twenty-first-century modes of capitalist subjectification and exploitation. Each of these accounts seeks to locate the possibility of practices or modes of individual and collective being that can evade and unmake the apparatuses of informatic control. They are disparate and often abstract, perhaps necessarily so, but it is important to note that each calls for activity rather than for indifference or acquiescence. One should be careful not to valorize the privileged possibility of "dropping out" or retreating. What is necessary is to make materialist attentiveness—not only to the composition of machines but also to the vast, uncountable specificities of sociality—into a weapon that can be used against the idealized, fuzzy metaphors and the specific, concrete forms of exploitation and dispossession that together constitute capitalism in the age of control.

Notes

Introduction

1. For an indicative sample of these periodizing texts, see Alain Touraine, *The Post-industrial Society: Tomorrow's Social History: Classes, Conflicts, and Culture in the Programmed Society* (New York: Random House, 1971); Daniel Bell, *The Coming of Post-industrial Society: A Venture in Social Forecasting* (New York: Basic Books, 1973); Marc Porat, *The Information Economy: Definition and Measurement* (Washington, DC: US Department of Commerce, 1977); Alvin Toffler, *The Third Wave* (New York: Bantam Books, 1980); Fredric Jameson, "Postmodernism, or, the Cultural Logic of Late Capitalism," *New Left Review* 146 (1984): 59–92; Ash Amin, *Post-Fordism: A Reader* (Cambridge, UK: Blackwell); Manuel Castells, *The Rise of the Network Society* (Cambridge, UK: Blackwell, 1996); Wendy Brown's work from "Neo-Liberalism and the End of Liberal Democracy," *Theory and Event* 7.1 (2003) to *Undoing the Demos* (Cambridge, MA: MIT Press, 2015); David Harvey, *A Brief History of Neoliberalism* (Oxford: Oxford University Press, 2005); Luc Boltanski and Ève Chiapello, *The New Spirit of Capitalism*, trans. Gregory Elliott (London: Verso, 2005); Michael Hardt and Antonio Negri, *Empire* (Cambridge, MA: Harvard University Press, 2000).

2. For these works, see note 1.

3. See Norbert Wiener, *Cybernetics, or Control and Communication in the Animal and the Machine* (1948; 7th ed., New York: Wiley, Paris: Herman et Cie, 1949). For accounts of the study and development of control systems before Wiener, see Stuart Bennett, *A History of Control Engineering 1800–1930* (London: Peter Peregrinus, 1979) and *A History of Control Engineering 1930–1955* (London: Peter Peregrinus, 1993); James R. Beniger, *The Control Revolution: Technological and Economic Origins of the Information Society* (Cambridge, MA: Harvard University Press, 1989); and David R. Mindell, *Between Human and Machine: Feedback, Control, and Computing before Cybernetics* (Baltimore: Johns Hopkins University Press, 2002).

4. For an account of the intersections of technology and governmentality in the terminology of control, see G. T. Guilbaud, *What Is Cybernetics?* trans. Valerie MacKay (London: Heinemann, 1959), 1–7.

5. Beniger, *The Control Revolution*, vi.

6. Ibid., 37–38.

7. Karl Deutsch, *The Nerves of Government: Models of Political Communication and Control* (1963; New York: Free Press, 1966), 76.

8. Tiqqun, "The Cybernetic Hypothesis" (in French), *Tiqqun 2* (2001), 53, 55. My thanks go to the anonymous translator(s) at tiqqun.jottit.com and Rachel Shapiro for the translation of passages from this article.

9. Bernard Stiegler, *For a New Critique of Political Economy*, trans. Daniel Ross (Cambridge, UK: Polity, 2010), 31–32. Stiegler goes on to show how grammatization accounts for recording systems such as writing and technical media as well as for the discretization and social reorganization of manual labor by way of machines:

> [W]ith the industrial revolution the process of grammatization constituting the history of mnemotechnics *suddenly surpasses the sphere of language* that is, also, the sphere of logos.... [T]he process of grammatization invests bodies. And in the first place, it discretizes the gestures of producers with the aim of making possible their automatic reproduction—while at the very same moment there also appear those machines and apparatuses for reproducing the visible and the audible that so caught the attention of Walter Benjamin, machines and apparatuses which grammatized perception and, through that, the affective activity of the nervous system. (32–33)

10. "The appearance of the Computational Universe at a moment in human history when computers have achieved unparalleled scope and importance is obviously not coincidental" (N. Katherine Hayles, *My Mother Was a Computer* [Chicago: University of Chicago Press, 2005], 3). Although the concept of the computable universe addressed by computer scientists and physicists such as Konrad Zuse, Edward Fredkin, and Stephen Wolfram clearly forms part of the historical logic addressed in this book, the focus here is on the conflation of computation (often in fuzzy, metaphorical form) and socioeconomic thought and practice under late capitalism rather than on fundamental questions of physical reality. A prototypical concept of the computable universe does occur in Charles Babbage's *Ninth Bridgewater Treatise* (1838); this concept of the universe is instructive when considering Babbage's intertwined work on political economy and computing machines and is thus addressed in chapter 1. For versions of the computable-universe thesis, see Konrad Zuse, *Rechnender Raum* (Braunschweig, Germany: Friedrich Vieweg & Sohn, 1969); Edward Fredkin, "Finite Nature," in *Proceedings of the XXVIIth Rencotre de Moriond* (Gif-sur-Yvette, FR: Editions Frontieres, 1992), and "Introduction to Digital Philosophy," *International Journal of Theoretical Physics* 42.2 (2003): 189–247; and Stephen Wolfram, *A New Kind of Science* (Champaign, IL: Wolfram Media, 2002).

11. Boltanski and Chiapello, *The New Spirit of Capitalism*, 104.

12. "[W]eil nur ist, was schaltbar ist" (Friedrich Kittler, "Real Time Analysis, Time Axis Manipulation" [in German], in *Draculas Vermächtnis: Technische Schriften* [Leipzig: Reclam, 1993], 192).

13. David Golumbia's concept of computationalism—a set of beliefs in the power and novelty of computation that allows it to pass for a radical agent of change while reinforcing multiple forms of instrumental reason and existing political and economic inequalities—is exemplary in tracking some of the ways in which this conceptual form of digitality is manifested. See David Golumbia, *The Cultural Logic of Computation* (Cambridge, MA: Harvard University Press, 2009).

14. See Alec H. Reeves, "Electric Signaling System," US Patent 2, 272, 070, filed November 22, 1939, and M. D. Fagen, ed., *A History of Engineering and Science in the Bell System*, vol. 2 (New York: Bell Telephone Laboratories, 1975), 316.

15. B. M. Oliver, J. R Pierce, and C. E. Shannon, "The Philosophy of PCM," *Proceedings of the Institute of Royal Engineers* 36.11 (1948): 1324.

16. W. M. Goodall, "Television by Pulse Code Modulation," *Bell Systems Technical Journal* 30.1 (1951): 33.

17. Ibid.

18. Oliver, Pierce, and Shannon, "The Philosophy of PCM."

19. Ibid., 1324.

20. Friedrich Kittler, "There Is No Software," in *Literature, Media, Information Systems* (Amsterdam: G&B Arts International, 1997), 152. In making this point, Kittler quotes the physicist Brosl Hasslacher at length:

[W]e use digital computers whose architecture is given to us in the form of a physical piece of machinery, with all its artificial constraints. We must reduce a continuous algorithmic description to one codable on a device whose fundamental operations are countable, and we do this by various forms of chopping into pieces, usually called discretization … The compiler then further reduces this model to a binary form determined largely by machine constraints.

The outcome is a discrete and synthetic microworld image of the original problem, whose structure is arbitrarily fixed by a differencing scheme and computational architecture chosen at random. The only remnant of the continuum is the use of radix arithmetic, which has the property of weighing bits unequally, and for nonlinear systems is the source of spurious singularities.

This is what we actually do when we compute up a model of the physical world with physical devices. This is not the idealized and serene process that we imagine when usually arguing about the fundamental structures of computation, and very far from Turing machines.

Kittler incorrectly cites Hasslacher's paper as "Algorithms in the World of Bounded Resources," but the correct title is "Beyond the Turing Machine" (quoted in "There Is No Software," 152). Yuri Gurevich's essay "Algorithms in the World of Bounded Resources" appears directly before Hasslacher's "Beyond the Turing Machine" in Rolf Herken, ed., *The Universal Turing Machine: A Half-Century Survey* (Oxford: Oxford University Press, 1988), 407–416 and 417–434.

21. See Peter Lunenfeld, *Snap to Grid: A User's Guide to Digital Arts, Media, and Cultures* (Cambridge, MA: MIT Press, 2001); Geert Lovink, *Dark Fiber* (Cambridge, MA: MIT Press, 2002); and Alexander R. Galloway and Eugene Thacker, *The Exploit: A Theory of Networks* (Minneapolis: University of Minnesota Press, 2007).

22. Lunenfeld, *Snap to Grid*, 174.

23. Fredric Jameson, "Class and Allegory in Contemporary Mass Culture: *Dog Day Afternoon* as a Political Film," *College English* 38.8 (1977): 846.

Chapter 1

1. See Gilles Deleuze, "Appendix: on the Death of Man and Superman," in *Foucault*, trans. Seán Hand (1986; reprint, London: Continuum, 1999), 102–110; "Having an Idea in Cinema," in Eleanor Kaufman and Kevin Jon Heller, eds., *Deleuze and Guattari: New Mappings in Politics, Philosophy, and Culture* (Minneapolis: University of Minnesota Press, 1998), 14–22; "Control and Becoming" and "Postscript on Control Societies," in *Negotiations*, trans. Martin Joughin (New York: Columbia University Press, 1995), 169–176 and 177–182.

2. Deleuze, "Postscript on Control Societies," 178.

3. Deleuze, "Control and Becoming," 175, emphasis added.

4. Deleuze, "Postscript on Control Societies," 180.

5. Ibid.

6. Deleuze, "Control and Becoming," 175.

7. Hardt and Negri, *Empire*, 23.

8. Alexander R. Galloway, *Protocol: How Control Exists after Decentralization* (Cambridge, MA: MIT Press, 2004), 3.

9. Wendy Hui Kyong Chun, *Control and Freedom: Power and Paranoia in the Age of Fiber Optics* (Cambridge, MA: MIT Press, 2006), 9.

10. Deleuze, "Control and Becoming," 175. For an exemplary account of communicative capitalism, see Christian Marazzi, *Capital and Affects: The Politics of the Language Economy* (1994; reprint, Cambridge, MA: MIT Press, 2011).

11. Deleuze, "Postscript on Control Societies," 178–179.

12. Castells, *The Rise of the Network Society*; Boltanski and Chiapello, *The New Spirit of Capitalism*. In *Protocol*, Galloway analyzes the connection between control societies and the network form at length.

13. Deleuze, "Postscript on Control Societies," 181.

14. Ibid., 179.

15. Ibid., 181–182. Earlier in "Postscript," Deleuze writes: "Even the state education system has been looking at the principle of 'getting paid for results': in fact, just as businesses are replacing factories, *school* is being replaced by *continuing education* and

exams by continuous assessment. It's the surest way of turning education into a business" (179).

16. Martin Joughin observes in a translator's note to "Postscript on Control Societies" that "the European Community/Union 'Exchange Rate Mechanism,' in which currencies are allowed to vary in value or 'float' within limitations set by their notional rate against a weighted basket of other participating currencies, was commonly called 'the snake'" ("Translator's Notes," in Deleuze, *Negotiations*, 203). Notably, Deleuze's metaphorical use of these two animals is distinct from that taken up by Hardt and Negri in *Empire*, which rests on Marx's depiction of revolutionary process as "methodical" like a mole in *The Eighteenth Brumaire of Louis Napoleon*. Although Hardt and Negri refer to Deleuze's control thesis elsewhere in *Empire*, they, following Marx, use the mole and the snake not to describe functions of money but to account for historically successive forms of revolutionary struggle (*Empire*, 57–58).

17. Deleuze, "Postscript on Control Societies," 181.

18. Jean-Louis Comolli, "Machines of the Visible," in Teresa De Lauretis and Steven Heath, eds., *The Cinematic Apparatus* (London: St. Martin's Press, 1980), 121.

19. For a nuanced discussion of the relationship between software and ideology, see Wendy Hui Kyong Chun, "On Software, or the Persistence of Visual Knowledge," *Grey Room* 18 (Winter 2004): 26–51, and Alexander R. Galloway's response, "Language Wants to Be Overlooked," *Journal of Visual Culture* 5.3 (2006): 315–331.

20. Deleuze, "Postscript on Control Societies," 180. Of course, many specific characteristics of late capitalism directly result from the possibilities afforded by computers and distributed digital networks, not least the management of highly dynamic and complex global systems of production and distribution in the retail and manufacture industries. Frederic Jameson notes in his analysis of Wal-Mart, for example, that

> it will be said that Wal-Mart may be a model of distribution but it can scarcely be said to be a model of production in the strict sense, however much we might talk of the production of distribution, etc. This cuts to the very heart of our socioeconomic contradictions: one face of which is structural unemployment, the other the definitive outstripping (dated in the US from 2003) of "productive" employment by retail employment. (Computerization and information would also have to be included in these new contradictory structures, and I think it is evident that Wal-Mart's special kind of success is dependent on computers and would have been impossible before them.) (*Valences of the Dialectic* [London: Verso, 2009], 423)

The present argument does not deny these obvious direct effects of computerization but rather suggests that the principle of (for example) a retail business conceived of as a system or network, in which individual workers are nodes that are functionally identical to material resources, money, patents, and so on, is also rooted in modes of socioeconomic thought and analytic logic that are not always directly mediated through electronic digital computers and that are in fact built on principles that predate such technologies.

21. Fredric Jameson, *The Political Unconscious: Literature as Socially Symbolic Act* (London: Routledge, 1983), 4.

22. The theorization of real subsumption as vaporizing communication and life developed by Negri, Paolo Virno, and others is a convincing and powerful one. However, as Neferti X. M. Tadiar shows, it fails to account for the parallel process through which some forms of life are deemed insufficient even for exploitation and are thus framed as superfluous. This book is motivated by Tadiar's observation to examine the interconnected technological and epistemic roots of immaterial labor in order to further show how such a conceptualization of labor is premised on exclusion. See Neferti X. M. Tadiar, "Life-Times in Fate-Playing," *South Atlantic Quarterly* 111.4 (2012): 783–802.

23. Deleuze, "Control and Becoming," 175.

24. Deleuze, "Appendix," 129–131. Here Deleuze describes the process through which life and labor, the subjects of the new and often interrelated disciplines of biology and political economy in the nineteenth century, become the subjects of informatic disciplines such as "molecular biology" and "cybernetics and information technology." Deleuze also writes of language, which becomes the subject of linguistics in the nineteenth century, but the relation of this category to Deleuze's broader periodization is confusing, focusing as it does on the emergence of the "agrammaticalities" of modern literature (Cummings, Burroughs, Roussel, Brisset, Dada collages) as the twentieth-century dispersal of linguistics, when it would appear to be more consistent with the other two categories to address computer programming or the Chomskian computationalist linguistics that Deleuze writes of together with Félix Guattari in *A Thousand Plateaus* (trans. Brian Massumi [Minneapolis: University of Minnesota Press, 1987]).

25. Stiegler, *For a New Critique*, 33–34.

26. Deleuze, "Postscript on Control Societies," 179.

27. Ibid.

28. Ibid.

29. Fredric Jameson, *Representing Capital: A Reading of Volume One* (London: Verso, 2011), 11–46. Or, according to Tiqqun, "The commodity is *essentially* that which is *absolutely equivalent*. This is revealed when two commodities (one of which is often money) are exchanged. Marx denounced this equivalence as an abstraction, and with reason: it is a *real* abstraction" ("On the Economy Considered as Black Magic" [in French], *Tiqqun* [1999]: 80, my translation).

30. Karl Marx, *Capital: A Critique of Political Economy*, vol. 1, trans. Ben Fowkes (London: Penguin Books, New Left Review, 1976), 128.

31. Ibid.

32. Lev Manovich, *The Language of New Media* (Cambridge, MA: MIT Press, 2001), 28.

33. Ernst Mandel, "Introduction," in Marx, *Capital*, 82.

34. Karl Marx, *Grundrisse*, trans. Ernst Wangerman, in *Karl Marx and Frederick Engels Collected Works*, vol. 28 (New York: International Publishers, 1986), 422.

35. Karl Marx, *Economic Manuscript of 1861–63*, trans. Ben Fowkes, in *Karl Marx and Frederick Engels Collected Works*, vol. 33 (London: Lawrence & Wishart, 1991), 387 (XIX.1169). All references to the *Economic Manuscript of 1861–63* in this text give page references for the *Collected Works* and then in parentheses the notebook and page numbers of the manuscript itself. For further accounts of the role of the clock in the formation of industrial disciplinary societies, see Lewis Mumford, *Technics and Civilization* (London: Routledge and Kegan Paul, 1967), and E. P. Thompson, "Time, Work-Discipline, and Industrial Capitalism," *Past and Present* 38 (December 1967): 56–97.

36. Michel Foucault, *Discipline and Punish*, trans. Alan Sheridan (London: Allen Lane, 1977), 157.

37. Karl Marx, *The Poverty of Philosophy*, trans. The Institute of Marxism-Leninism, in *Karl Marx and Frederick Engels Collected Works*, vol. 6 (Moscow: Progress, 1976.), 127.

38. Karl Marx, *Economic and Philosophic Manuscripts of 1844*, trans. Martin Milligan and Dirk J. Struik, in *Karl Marx and Frederick Engels Collected Works*, vol. 3 (London: Lawrence & Wishart, 1975), 274.

39. Ibid.

40. Ibid.

41. György Lukács, *History and Class Consciousness* (1967; reprint, London: Merlin Press, 1971), 90–91.

42. Deleuze and Guattari, "Apparatus of Capture," in *A Thousand Plateaus*, 468–522; Giorgio Agamben, *What Is an Apparatus? and Other Essays*, trans. David Kishik and Stefan Pedatella (Stanford, CA: Stanford University Press, 2009), 14.

43. Philip E. Agre, "Surveillance and Capture: Two Models of Privacy," *Information Society* 10.2 (1994): 101.

44. Philip E. Agre, "Beyond the Mirror World: Privacy and the Representational Practices of Computing," in Philip E. Agre and Marc Rotenberg, eds., *Technology and Privacy: The New Landscape* (Cambridge, MA: MIT Press, 1997), 32.

45. Gary Becker, *The Economic Approach to Human Behavior* (Chicago: University of Chicago Press, 1976), 5. The isomorphism between this basic approach to societies and the approach posited in structuralist anthropology (itself a discipline closely

related to certain deployments of computational metaphors) is marked and is discussed further in chapter 2.

46. Ibid.

47. Michel Foucault, *The Birth of Biopolitics: Lectures at the Collège de France 1978–79*, trans. Graham Burchell (London: Palgrave Macmillan, 2008), 219.

48. Ibid., 225.

49. Becker, *The Economic Approach to Human Behavior*, 7.

50. Foucault, *The Birth of Biopolitics*, 223, 225.

51. Friedrich Hayek, "The Use of Knowledge in Society," *American Economic Review* 35.4 (1945): 527.

52. Maurizio Lazzarato, "Immaterial Labor," trans. Paul Colilli and Ed Emory, in Michael Hardt and Paolo Virno, eds., *Radical Thought in Italy: A Potential Politics* (Minneapolis: University of Minnesota Press, 1996), 133.

53. Ibid.

54. Maurizio Lazzarato, "From Capital-Labor to Capital-Life," *Ephemera* 4.3 (2004): 188.

55. Paolo Virno, *A Grammar of the Multitude: For an Analysis of Contemporary Forms of Life*, trans. Isabella Bertoletti, James Cascaito, and Andrea Casson (Los Angeles: Semiotext(e), 2004), 50.

56. Jonathan Beller, *The Cinematic Mode of Production: Attention Economy and the Society of the Spectacle* (Hanover, NH: Dartmouth College Press, 2006), 76, italics added.

57. Karl Marx, *Economic Manuscript of 1861–63*, 387 (XIX.1159).

58. *The Industry of Nations Part II: A Survey of the Existing State of Arts, Machines, and Manufactures* (London: Society for Promoting Christian Knowledge), 159–160, 162, quoted in Marx, *Economic Manuscript of 1861–63*, 411 (XIX.1177).

59. *The Industry of Nations*, 160–161, quoted in Marx, *Economic Manuscript of 1861–63*, 411 (XIX.1177). Among many other things, this section of *The Industry of Nations* is remarkable for the way in which it posits a protocomputational system (the Jacquard loom) as a precursor to the (supposedly precomputational) mechanical reproduction of art that Walter Benjamin wrote of in 1936. It says:

It might be supposed that no arrangement of this kind could produce any very delicate results, and that the patterns must be coarse and clumsy in their details. But to so great a nicety have machinists now brought this beautiful apparatus, and so accurately are the perforated cards arranged so as to fulfil their task of operating or ceasing to operate upon the healds [*sic*], that the most delicate a beautiful designs are now accomplished by its means. Engravings have been copied with minute accuracy, and portraits have been embroidered with so much delicacy and vigour as almost to equal fine engravings on paper. At the Great Exhibition were shown by a

French weaver, a portrait of the Pope, another of the Duc d'Aumale, and another of the Queen of England, all produced by the Jacquard apparatus and executed with artistic skill of a high order. (162)

60. Marx, *Capital*, 466, 469, 497, 514, 528–529.

61. Marx, *The Poverty of Philosophy*, 155; *Economic Manuscripts of 1861–1863*, 388 (XIX.1160); *Capital*, 497 n. 10.

62. Charles Babbage, *On the Economy of Machinery and Manufactures* (London: Charles Knight, 1932), 136. It appears that Marx quotes from the 1833 French translation of Babbage's text, and minor differences in translation are found in each of the editions cited in this book.

63. For an efficient account of Marx's account of machine as sociotechnical concatenation and the relationship of this concatenation both to concepts of the machine in ancient Greece and Rome and to a range of twentieth-century critical-theoretical concepts, including the machinic in Felix Guattari and Gilles Deleuze, see Gerald Raunig, *A Thousand Machines: A Concise Philosophy of the Machine as Social Movement* (Los Angeles: Semiotext(e), 2010).

64. Babbage, *On the Economy of Machinery*, iii.

65. Philip Mirowski, *Machine Dreams: Economics Becomes a Cyborg Science* (Cambridge: Cambridge University Press, 2002), 33.

66. Ibid., 34.

67. Babbage, *On the Economy of Machinery*, 153.

68. Ibid., 162.

69. Charles Babbage, *The Ninth Bridgewater Treatise: A Fragment* (London: John Murray, 1837), 33–34.

70. Anthony Hyman, *Charles Babbage: Pioneer of the Computer* (Oxford: Oxford University Press, 1982), 138–139.

71. Rowland Hill and George Birkbeck Hill, *The Life of Rowland Hill and the History of Penny Postage*, vol. 1 (London: Thomas De La Rue, 1880), 112–113. Both Bruce Castle and the Hazlewood school were run as experiments according to principles developed by Rowland Hill, who would go on to invent the penny post, and his brother Matthew. See Rowland Hill and Matthew Hill, *Plans for the Government and Liberal Instruction of Boys in Large Numbers; Drawn from Experience in the Hazlewood School* (London: G. and W. B. Whittaker, 1822). On Babbage's 1827 visit to Hazlewood and his subsequent correspondence with Rowland and Matthew Hill on calculating engines, education, and postal reform, see Hyman, *Charles Babbage*, 64–65, and Bernhard Siegert, *Relays: Literature as an Epoch of the Postal System*, trans. Kevin Repp (Stanford, CA: Stanford University Press, 1999), 122–125.

72. Hill and Hill, *Plans for the Government and Liberal Instruction of Boys*, 107–108; Siegert, *Relays*, 123.

73. Hill and Hill, *Plans for the Government and Liberal Instruction of Boys*, 28.

74. Tara McPherson, "U.S Operating Systems at Mid-century: The Intertwining of Race and UNIX," in Lisa Nakamura and Peter A. Chow-White, eds., *Race after the Internet* (New York: Routledge, 2011), 24.

75. Jonathan Beller, "The Martial Art of 'Cinema': Modes of Virtuosity á la Hong Kong and the Philippines," *Positions* 19.2 (2011): 281.

76. For a detailed account of Hollerith's method as well as of the subsequent development of punch-card tabulation of U.S. Census data, see Leon E. Truesdell, *The Development of Punch Card Tabulation in the Bureau of the Census 1890–1940* (Washington, DC: US Department of Commerce, 1965), esp. 26–82.

77. Herman Hollerith, "The Electrical Tabulating Machine," *Journal of the Royal Statistical Society* 57.4 (1894): 678–679.

78. Herman Hollerith, "An Electrical Tabulating System," *Columbia University School of Mines Quarterly* 10.16 (1889): 252–253.

79. Ibid., 241.

80. Ibid., 242.

81. Ibid., 241–242.

82. Ibid., 241.

83. Wiener, *Cybernetics*, 34, emphasis in original.

84. Ibid.

85. Vilém Flusser, *Toward a Philosophy of Photography* (London: Reaktion Books, 2000), 21.

86. Although Foucault's account of the emergence of techniques of biopolitical management is rooted in the management of individuated bodies, it does clearly hint at the informatic modes of atomized control that were emerging at the time he was writing. Foucault's depiction of biopolitics as striving toward "homeostasis," "equilibrium," and "regulation," "bring[ing] together the mass effects characteristic of a population," "control[ling] ... series of random events," and "predict[ing] the probability of those events" is clearly redolent of cybernetic terminology and techniques (*"Society Must be Defended": Lectures at the Collège de France, 1975–76*, trans. David Macey [New York: Picador, 2003], 246, 249). This is doubtless why Deleuze observes that Foucault was "one of the first to say that we're moving away from disciplinary societies, that we've left them behind" ("Control and Becoming," 174).

87. See, for example, Martin Heidegger, "The End of Philosophy and the Task of Thinking," trans. Joan Stambaugh, in *Basic Writings*, ed. David Farrell Krell (Oxford: Routledge, 1993), 373–392; Jean-François Lyotard, *The Postmodern Condition: A Report on Knowledge*, trans. Geoffrey Bennington and Brian Massumi (Manchester, UK: Manchester University Press, 1986); N. Katherine Hayles, *How We Became Posthuman: Virtual Bodies in Cybernetics, Literature, and Informatics* (Chicago: University of Chicago Press, 1999); Steve Joshua Heims, *The Cybernetics Group, 1946–1953: Constructing a Social Science for Postwar America* (Cambridge, MA: MIT Press, 1991); Andrew Pickering, *The Cybernetic Brain: Sketches of Another Future* (Chicago: University of Chicago Press, 2011). Notable counterexamples can be found in Tiqqun, "The Cybernetic Hypothesis," and Mirowski, *Machine Dreams*. Lazzarato names cybernetics as central to the concept of immaterial labor but uses the term to describe the use of computers and automated machinery.

88. See Frederick Lanchester, *Aircraft in Warfare: The Dawn of the Fourth Arm* (London: Constable, 1916).

89. Nancy Fraser, "Feminism, Capitalism and the Cunning of History," *New Left Review* 56 (2009): 97–117.

Chapter 2

1. Peter Galison, "War against the Center," *Grey Room* 4 (Summer 2001): 7, 8.

2. Ibid., 8.

3. Ibid., 8–9

4. Ibid., 12.

5. Ibid., 29, 30.

6. Ibid., 29, emphasis added. Galison has written elsewhere on the historical process through which the subject of early cybernetic research shifted from enemy to ally. See Peter Galison, "The Ontology of the Enemy: Norbert Wiener and the Cybernetic Vision," *Critical Inquiry* 21.1 (1994): 228–266.

7. See Wiener's introduction to *Cybernetics* as well as Steve Joshua Heims, *John von Neumann and Norbert Wiener: From Mathematics to the Technologies of Life and Death* (Cambridge, MA: MIT Press, 1980), and *The Cybernetics Group*; Galison, "The Ontology of the Enemy"; and Mindell, *Between Human and Machine*.

8. See Galison, "The Ontology of the Enemy," 231–232. Galison adopts the term *Manichean* to describe these sciences of predictive management from Wiener's characterization of the "Manichean Devil," who is "determined on victory and will use any trick of craftiness or dissimulation to obtain this victory," in contrast to the "Augustinian Devil" (such as forces of nature), which, in Galison's paraphrasing of

Wiener, "was characterized by the 'evil' of chance or disorder but could not change the rules" (231–232).

9. The cybernetic hypothesis is not limited to formal cybernetic practices but is more broadly "a political hypothesis" that "proposes to conceive of biological, physical, and social behaviors as ... integrally programmed and reprogrammable" and that "conceives of each individual behavior as something 'piloted,' in the last analysis, by the need for the survival of a 'system' that makes it possible, and which it must contribute to" (Tiqqun, "The Cybernetic Hypothesis," 42).

10. Wiener, *Cybernetics*, 19–20.

11. Ibid., 9.

12. It should be noted that the prediction written of here is probabilistic, in contrast to the deterministic ideas that characterized earlier mechanistic views of human and universe.

13. Oliver, Pierce, and Shannon, "The Philosophy of PCM," 1324.

14. Alan Turing, "Computing Machinery and Intelligence," *Mind* 59.236 (1950): 440.

15. Arturo Rosenblueth, Norbert Wiener, and Julian Bigelow, "Behavior, Purpose, and Teleology," *Philosophy of Science* 10.1 (1943): 19–20.

16. Wiener, *Cybernetics*, 11.

17. Ibid., 12.

18. As argued in this chapter, these limitations do not absolve Wiener of playing a major role in the emergence of the control stage of capitalism. The basic principles of statistical prediction applied to complex systems can be seen to consolidate around his work in this period. As Friedrich Kittler writes, "With Wiener's Linear Prediction Code (LPC), mathematics changed into an oracle capable of predicting a probable future even out of chaos initially for fighter aircraft and anti-aircraft guidance systems, in between the wars for human mouths and the computer simulations of their discourses" (*Gramophone, Film, Typewriter*, trans. Geoffrey Winthrop-Young and Michael Wutz [Stanford, CA: Stanford University Press, 1999], 260).

19. Wiener, *Cybernetics*, 15–16.

20. Ibid., 16.

21. Ibid., 16–17.

22. Ibid., 17.

23. Ibid., 19.

24. Participants in this meeting included anthropologists Gregory Bateson and Margaret Mead; interdisciplinary social scientist and behaviorist Lawrence K. Frank; psychobiologist and behaviorist Howard Liddell; psychiatrists Milton H. Erickson, Lawrence S. Kubie, and Warren McCulloch; and physiologist Arturo Rosenblueth. See Heims, *The Cybernetics Group*, 14–17. On the connections between work carried out in this period and the development of cognitive science, see Jean-Pierre Dupuy, *On the Origins of Cognitive Science: The Mechanization of the Mind*, trans. M. B. DeBevoise (Cambridge, MA: MIT Press, 2009).

25. Warren McCulloch and Walter Pitts, "A Logical Calculus of the Ideas Immanent in Nervous Activity," *Bulletin of Mathematical Biophysics* 5 (1943): 129; Alan Turing, "On Computable Numbers with an Application to the *Entscheidungsproblem*," *Proceedings of the London Mathematical Society* Series 2, 42 (1936): 230–265.

26. McCulloch and Pitts, "A Logical Calculus," 132.

27. Warren McCulloch, "An Account of the First Three Conferences on Teleological Mechanisms," memo addressed "To the Members of the Fourth Conference on Teleological Mechanisms," reproduced in Claus Pias, ed., *Cybernetics—Kybernetik: The Macy Conferences 1946–1953* (Zurich: Diaphenes, 2004), 337–339.

28. Ibid., 339.

29. Ibid.

30. This break between the physiologists, mathematicians, and engineers on the one hand and the social scientists on the other was so marked that the "smoldering" of the first Macy conference led to the organization of a special meeting for the social scientists in September 1946. Bateson organized this event, and in addition to Wiener and von Neumann its attendees included Talcott Parsons, Robert K. Merton, and Clyde Kluckhohn. See McCulloch, "An Account of the First Three Conferences," 339.

31. Ibid., 338. The article "Bali: The Value System of a Steady State" was first published in Meyer Fortres, ed., *Social Structure: Studies Presented to A. R. Radcliffe-Brown* (New York: Russell & Russell, 1963), and is reprinted in Gregory Bateson, *Steps to an Ecology of Mind: Collected Essays in Anthropology, Psychiatry, Evolution, and Epistemology* (Chicago: University of Chicago Press, 2000), 107–127.

32. Rosenblueth, Wiener, and Bigelow, "Behavior, Purpose, and Teleology," 20; see also McCulloch, "An Account of the First Three Conferences."

33. Dupuy, *On the Origins of Cognitive Science*, 122.

34. Ibid.

35. Deleuze and Guattari, *A Thousand Plateaus*, 352.

36. This conversation is reported in William Poundstone, *Prisoner's Dilemma* (New York: Anchor Books, 1993), 6.

37. Von Neumann's work on game theory began some twenty years before the publication of the book he coauthored with Oskar Morgenstern, *Theory of Games and Economic Behavior* (Princeton, NJ: Princeton University Press, 1944), but it is this book that inaugurated the expansion of game theory as a major influence on economic and military thought from the 1940s on. Von Neumann's earliest publication on the subject is "Zur Theorie der Gesellschaftsspiele," *Mathematische Annalen* 100.1 (1928). The literature on game theory and its uses is extensive, and it is impossible to give a thorough account here. For overviews from the first two decades after the publication of *Theory of Games and Economic Behavior*, see John McDonald, *Strategy in Poker, Business, and War* (New York: Norton, 1950), and Deutsch, *The Nerves of Government*, 51–71. For examples of the continued reach of game theory into political and economic thought after the 1970s, see Martin Shubik, *Game-Theoretic Approach to Political Economy* (Cambridge, MA: MIT Press, 1987), and Roger Myerson, *Game Theory: Analysis of Conflict* (Cambridge, MA: Harvard University Press, 1991).

38. John von Neumann and Oskar Morgenstern, *Theory of Games and Economic Behavior*, 3rd ed. (Princeton, NJ: Princeton University Press, 1953), 3.

39. Wiener, *Cybernetics*, 186. On the impossibility of prediction, Wiener goes on to state that "instead of looking after his own ultimate interest, after the fashion of von Neumann's gamesters, the fool operates in a manner which, by and large, is as predictable as the struggles of a rat in a maze" (186).

40. Gregory Bateson to Norbert Wiener, September 22, 1952, quoted in Heims, *John von Neumann and Norbert Wiener*, 307–308.

41. Ibid., quoted in Mirowski, *Machine Dreams*, 65.

42. Bateson, "Bali," 125.

43. Ibid., 126.

44. Maya Deren, "From the Notebook of Maya Deren, 1947," *October* 14 (Autumn 1980): 23, 40. For a longer discussion of Deren's response to Bateson and its relationship to her filmmaking practice, see Orit Halpern, "Anagram, Gestalt, Game in Maya Deren: Reconfiguring the Image in Post-war Cinema," *Postmodern Culture* 19.3 (2009): http://muse.jhu.edu/journals/postmodern_culture/v019/19.3.halpern.html, accessed February 8, 2015.

45. Deren writes in the notebook entry dated March 16, 1947, that "man is distinguished for consciousness, time perspective, and original energy.... Consequently, it would seem to me that any effort to pattern a society should take its cue from that distinctive feature in man which is unlike machines, and which makes him capable of building machines" ("From the Notebook of Maya Deren, 1947," 40). As Halpern puts it, Deren "interrogates the universalist assumptions of these models" and

"intuitively ... identifies a homogenizing force within the technical logic of the game" ("Anagram, Gestalt, Game").

46. Wiener, *Cybernetics*, 38.

47. Mirowski, *Machine Dreams*, 139. Although the focus here is on the confluence of cybernetic thought and economism, it is worth noting that the movement from game theory to the development of computing machines and wider-ranging theories of complex systems broadly coincides with von Neumann's attachment to ordnance research, which ultimately led to his role in the atomic bomb project at Los Alamos. See Mirowski, *Machine Dreams* 136–137.

48. Wiener, *Cybernetics*, 50. Each of these mechanistic metaphors is, of course, imbricated with the progressive expansion and intensification of capitalism. Silvia Federici perhaps puts this best when she argues of the seventeenth-century reformulations of the human as mechanism that "the outcome is a redefinition of bodily attributes that makes the body, ideally, at least, suited for the regularity and automatism demanded by the capitalist work-discipline" (*Caliban and the Witch* [New York: Autonomedia, 2004], 138).

49. Deleuze, "Postscript on Control Societies," 180.

50. Lawrence Kubie, "The Fallacious Use of Quantitative Concepts in Dynamic Psychology," *Psychoanalytic Quarterly* 16 (1947): 516–518. For a lengthier discussion of the historical relationship between cybernetics and psychoanalysis, see chapter 5 in this book.

51. Norbert Wiener, *The Human Use of Human Beings: Cybernetics and Society* (London: Free Association Books, 1989), 11. These energetic metaphors are explicitly connected to nineteenth-century technical standards by a number of other writers; see, for example, Michel Serres, *Hermes: Literature, Science, Philosophy*, ed. Josue V. Harari and David F. Bell (Baltimore, MD: Johns Hopkins University Press, 1982), 72.

52. Norbert Wiener, "Some Moral and Technical Consequences of Automation," *Science* 131 (May 1960): 1358.

53. John von Neumann, "First Draft of a Report on the EDVAC," contract no. W-670-ORD-4926 between the US Army Ordnance Department and the University of Pennsylvania (June 1945), sec. 4.1.

54. Ibid.

55. Ibid., sec. 4.2.

56. Ibid.

57. Ibid., sec. 5.4, 5.6, italics in original.

58. David A. Patterson and John L. Hennessy, *Computer Organization and Design: The Hardware/Software Interface* (San Francisco: Morgan Kaufmann, 1998), 59, italics in original.

59. Ibid.

60. Mirowski, *Machine Dreams*, 146, 147.

61. A slightly edited version of this talk is published as "The General and Logical Theory of Automata," in *John von Neumann Collected Works*, vol. 5: *Design of Computers, Theory of Automata, and Numerical Analysis*, ed. A. H. Taub (Oxford: Pergamon Press, 1963), 288–326.

62. Von Neumann, "The General and Logical Theory of Automata," 288–289.

63. As Wendy Hui Kyong Chun observes, "The General and Logical Theory of Automata" reverses the foundational analogy of "First Draft of a Report on the EDVAC." Where in the earlier "First Draft" the ideal neuron provides a model for thinking about the digital machine, in the later essay the ideal computing machine provides an analogy for the nervous system (*Programmed Visions: Software and Memory* [Cambridge, MA: MIT Press, 2011], 157–158).

64. Von Neumann, "The General and Logical Theory of Automata," 296.

65. Arthur Burks, editorial note in John von Neumann, *Theory of Self Reproducing Automata* (Urbana: University of Illinois Press, 1966), 21. The assumption that economic systems are "natural" is indicative of the difference between the broadly opposed camps in the early years of cybernetics research.

66. Mirowski evokes this world picture in a passage that is worth quoting at length:

The von Neumann armada can seem gunmetal grey, sitting distressingly low in the water but dispersed from here to eternity, and more than a little forbidding to human-centric concerns. It is not centrally engaged with psychology or neurophysiology, and therefore is not really occupied in building up the architecture of societies from individual beliefs and intentions. It is not even concerned with getting any one agent "right" in any particular simulation, because the individual agent will not bulk large in the Neumannesque scheme of things. Rather, it is the root theory of abstract information processors and their interactions, ranging all the way from the genome to the market.... Casting causal entities as machines is standard modus operandi; erasing boundaries between humans and machines hardly raises an eyebrow. (*Machine Dreams*, 470–471)

67. F. S. C. Northrop, "The Neurological and Behavioristic Psychological Basis of the Ordering of Society by Means of Ideas," *Science* 107 (April 1948): 416.

68. See McCulloch, "An Account of the First Three Conferences."

69. Northrop, "Neurological and Behavioristic Psychological Basis," 416.

70. Claude Lévi-Strauss, "Language and the Analysis of Social Laws," in *Structural Anthropology*, trans. Claire Jacobson and Brooke Grundfest Schoepf (New York: Basic Books, 1963), 56. For the 1949 presentation of this essay within a compelling and rigorous account of Lévi-Strauss's discovery of and interest in cybernetics, see Bernard Dionysius Geoghegan, "From Information Theory to French Theory: Jakobson, Lévi-Strauss, and the Cybernetic Apparatus," *Critical Inquiry* 38 (Autumn 2011): 116 n. 56; 116 n. 58.

71. Lévi-Strauss, "Language and the Analysis of Social Laws," 58–59.

72. Ibid., 59.

73. Ibid.

74. Lazzarato, "Immaterial Labor," 133.

75. Lévi-Strauss, "Language and the Analysis of Social Laws," 61. For a wider discussion of demographic modeling after cybernetics, see chapter 3 of the present book.

76. Ibid.

77. Mirowski, *Machine Dreams*, 208.

78. Robert F. Bales, "Social Interaction," RAND Paper P-587, second draft, December 14, 1954, 1–2.

79. Ibid., 1.

80. Ibid., 3.

81. Ibid., 10.

82. Ibid., 5.

83. Ibid., 6.

84. Dupuy, *On the Origins of Cognitive Science*, 61.

85. For lengthy discussions of the latter two developments, see Mirowski, *Machine Dreams*, 232–308, 319–369.

86. W. Ross Ashby, "The Effect of Controls on Stability," *Nature* 155 (February 24, 1945), 242, 243.

87. W. Ross Ashby, *Journal* (1928–1972), W. Ross Ashby Digital Archive, http://www.rossashby.info/journal/page/1799+06.html, accessed February 7, 2014.

88. Ibid. In a letter to Ashby dated December 4, 1946, Bateson states that the point Ashby made in the *Nature* letter (cited in note 86) is "of the very highest importance," but he reveals a quite different political orientation (and, indeed, a quite different economic potential for cybernetics) by stating that it would be of primary interest to a "government desiring to control some economic variable" rather than to the advocate of markets as regulatory systems (Gregory Bateson to W. Ross Ashby, December 4, 1946, W. Ross Ashby Digital Archive, http://www.rossashby.info/letters/bateson.html, accessed February 7, 2014). For an account of computation and central planning, see Nick Dyer-Witheford, "Red Plenty Platforms," *Culture Machine* 14 (2013), http://www.culturemachine.net/index.php/cm/article/viewArticle/511 (accessed February 5, 2015), and the material on Anthony Stafford Beer in this chapter.

89. See Dupuy, *On the Origins of Cognitive Science*, 140, and Mirowski, *Machine Dreams*, 235. In an unpublished work from the mid-1950s, Hayek demonstrates a familiarity with "cybernetics, the theory of automata or machines, general systems theory and ... communication theory" (Bruce Caldwell, *Hayek's Challenge: An Intellectual Biography of F.A. Hayek* [Chicago: University of Chicago Press, 2004], 303).

90. Friedrich Hayek, "The Sensory Order After 25 Years," in Walter B. Weimer and David S. Palermo, eds., *Cognition and the Symbolic Process* (Hillsdale, NJ: Erlbaum, 1982), 32.

91. Mirowski, *Machine Dreams*, 238.

92. Ibid., 518.

93. It is not possible to give a thorough account of Beer's work here. For an overview of his work and its relationship to British cybernetics, see Andrew Pickering, *The Cybernetic Brain: Sketches of Another Future* (Chicago: University of Chicago Press, 2010), esp. chap. 6.

94. Anthony Stafford Beer, *Cybernetics and Management* (1959; 2nd ed., London: English University Press, 1967), 142, 143.

95. This address is published in Anthony Stafford Beer, *Platform for Change: A Message from Stafford Beer* (London: Wiley, 1975), 23–37.

96. Ibid., 32, 36, 37.

97. Warren McCulloch, "Recollections of the Many Sources of Cybernetics," *ASC Forum* 6.2 (1974): 11.

98. See Eden Medina, *Cybernetic Revolutionaries: Technology and Politics in Allende's Chile* (Cambridge, MA: MIT Press, 2011), with the story of Beer's socialism on page 41. For a different account of the way in which cybernetics appealed to noncapitalist worldviews, see Slava Gerovitch, *From Newspeak to Cyberspeak: A History of Soviet Cybernetics* (Cambridge, MA: MIT Press, 2002). An interesting semifictional report of Soviet cybernetics can be found in Francis Spufford, *Red Plenty* (London: Faber and Faber, 2011).

99. Medina, *Cybernetic Revolutionaries*, 249 n. 17.

100. Norbert Wiener to W. Ross Ashby, April 8, 1952, W. Ross Ashby Digital Archive, http://www.rossashby.info/letters/wiener.html, accessed February 7, 2014.

101. Although neither Forrester nor Wiener cites the other, it is likely that they would have been at least aware of each other through MIT.

102. Like Bigelow, Shannon, Wiener, and von Neumann, Forrester worked on ballistics systems during World War II and worked on the SAGE (Semi Automatic Ground Environment) missile defense system between 1949 and 1956. For a detailed

discussion of the latter, see Paul N. Edwards, *The Closed World: Computer and the Politics of Discourse in Cold War America* (Cambridge, MA: MIT Press, 1996), 75–111.

103. Jay Wright Forrester, *Industrial Dynamics* (Cambridge, MA: MIT Press, 1961), 94.

104. Hardt and Negri, *Empire*, 290.

105. Beer, *Cybernetics and Management*, 145; Lazzarato, "Immaterial Labor," 135.

106. Foucault writes that as a first principle of biopolitical management, "regulatory mechanisms must be established to establish an equilibrium, maintain an average, establish a sort of homeostasis, and compensate for variations within this general populating and its aleatory field" (*Society Must Be Defended*, 246).

107. Peter Wegner, "Why Interaction Is More Important Than Algorithms," *Communications of the ACM* 40.5 (1997): 83.

108. Boltanski and Chiapello, *The New Spirit of Capitalism*, 155.

Part II

1. Neferti X. M. Tadiar, "Life-Times of Becoming Human," *Occasion: Interdisciplinary Studies in the Humanities* 3 (March 2012): 3.

2. Tiqqun, "The Cybernetic Hypothesis," 50, italics in original.

3. Geert Lovink and Florian Schneider, "Notes on the State of Networking," *Makeworlds* 4 (2004), http://makeworlds.net/node/100, accessed February 9, 2014.

Chapter 3

1. Tadiar, "Life-Times in Fate-Playing," 786–787.

2. Ibid., 787.

3. Ibid.

4. Ibid.

5. Ibid., 788. At the end of this passage, Tadiar quotes Aaron Benanav, "Misery and Debt: On the Logic and History of Surplus Populations and Surplus Capital," *Endnotes*, no. 2 (2010), http://endnotes.org.uk/articles/1, accessed February 7, 2014.

6. Saskia Sassen, *Expulsions: Brutality and Complexity in the Global Economy* (Cambridge, MA: Belknap Press of Harvard University Press, 2014), 36.

7. Kittler, "There Is No Software," 152.

8. Philipp von Hilgers casts doubt over the date of this meeting. Both Wiener and McCulloch (with the latter appearing to rely on the former's account) state that it took place in winter 1943–1944, whereas von Hilgers states that it took place on January 6–7, 1945 ("The History of the Black Box: The Clash of a Thing and Its Concept," *Cultural Politics* 7.1 [2011], 55 n. 13).

9. McCulloch, "Recollections," 12.

10. Von Neumann, "General and Logical Theory of Automata," 289.

11. Ibid., 313. The other appearances of the black-box concept in "General and Logical Theory of Automata" are on pages 293, 298–299, and 309.

12. William Ross Ashby, *An Introduction to Cybernetics*, 2nd ed. (London: Chapman and Hall, 1957), 6.

13. William Ross Ashby, *Design for a Brain* (New York: Wiley, 1954), 9, 32.

14. Ibid., 10.

15. Ibid., 11.

16. Hans-Jörg Rheinberger, *Toward a History of Epistemic Things: Synthesizing Proteins in the Test Tube* (Stanford, CA: Stanford University Press, 1997), 28.

17. See Paul Baran, "On Distributed Communications I: Introduction to Distributed Communications Networks," RAND Memorandum RM-3420-PR, August 1964.

18. The literature on networked and system-dynamic approaches to military strategy is extensive. For the canonical examples of the former, see John Arquilla and David Ronfeldt, *The Advent of Netwar* (Santa Monica, CA: RAND, 1996), and *Networks and Netwars: The Future of Terror, Crime, and Militancy* (Santa Monica, CA: RAND, 2001). For examples of the latter, see the essays in Reiner K. Huber, ed., *Systems Analysis and Modeling in Defense* (Oxford: Pergamon Press, 1984).

19. This body of work, along with the military-strategic approaches cited in the previous note, sits within the tradition of operations research and system dynamics analyzed in the previous chapter. See, for example, Forrester, *Industrial Dynamics*, and Stafford Beer, *Brain of the Firm* (1972; 2nd ed., New York: Wiley, 1995). For a more recent example attesting to the influence of these approaches on contemporary economic practices, see John D. Sterman, *Business Dynamics: Systems Thinking and Modeling for a Complex World* (New York: McGraw-Hill, 2000).

20. Geoff Mulgan, *Communication and Control* (Cambridge, UK: Polity, 1991) and *Connexity* (Boston, MA: Harvard Business Review Press, 1997). *Adding it Up: Improving Analysis and Modelling in Central Government* (London: Cabinet Office, 2001), 3.

21. See Bruno Latour, *Science in Action: How to Follow Scientists and Engineers through Society* (Milton Keynes, UK: Open University Press, 1987); Castells, *The Rise of the Network Society*; Albert-László Barabási, *Linked: The New Science of Networks* (Cam-

bridge: Perseus, 2002); Yochai Benkler, *The Wealth of Networks: How Social Production Transforms Markets and Freedom* (New Haven, CT: Yale University Press, 2006); and Franco Moretti, "Network Theory, Plot Analysis," *New Left Review* 68 (March–April 2011): 80–102.

22. Jameson specifically draws a parallel between Kevin Lynch's "conception of city experience" as "the dialectic between the here and now of immediate perception and the imaginative or imaginary sense of the city as an absent totality," on the one hand, and Althusser's famous formulation of ideology as "the Imaginary representation of the subject's relationship to his or her Real conditions of existence," on the other ("Cognitive Mapping," in Cary Nelson and Lawrence Grossberg, eds., *Marxism and the Interpretation of Culture* [Urbana: University of Illinois Press, 1988], 353).

23. Fredric Jameson, *The Geopolitical Aesthetic: Cinema and Space in the World System* (Bloomington: Indiana University Press; London: BFI, 1992), 2.

24. Chun, *Programmed Visions*, 71, 8.

25. As Jameson puts it, "Our faulty representations of some immense communicational and computer network are themselves but a distorted figuration of something even deeper, namely the whole world system of present-day multinational capitalism. The technology of contemporary society is therefore mesmerizing and fascinating, not so much in its own right, but because it seems to offer some privileged representational shorthand for grasping a network of power and control even more difficult for our minds and imaginations to grasp—namely the whole new decentred global network of the third stage of capital itself" ("Postmodernism," 79–80).

26. McCulloch, "Recollections," 16; Tiqqun, "The Cybernetic Hypothesis," 49.

27. Alexander R. Galloway, "Are Some Things Unrepresentable?" in *The Interface Effect* (Cambridge, UK: Polity, 2012), 98–99.

28. Ibid., 99.

29. Alexander R. Galloway, *Gaming: Essays on Algorithmic Culture* (Minneapolis: University of Minnesota Press, 2006), 103.

30. Ibid.

31. For a theorization of the necessity of affective (rather than cognitive) mapping to engage with control-era cultural forms, see Steven Shaviro, *Post-Cinematic Affect* (Winchester, UK: 0 Books, 2010).

32. Agamben, *What Is an Apparatus* 21, emphasis added.

33. Jeff De Joseph, "Beyond Knowing," *Adweek*, September 1999. The text of this article appeared in Bernadette Corporation's exhibit *2000 Wasted Years: Corporate Responsibility and the Swine We Are* at the Institute of Contemporary Arts, London, March 27–June 9, 2013.

34. Golumbia, *The Cultural Logic of Computation*, 131–133. In referring to these groups and segments, Golumbia cites the Claritas Corporation's website at http://www.claritas.com.

35. Urban Outfitters Inc. Analyst/Investor Day, September 27, 2012, transcript at http://seekingalpha.com/article/893321-urban-outfitters-inc-analyst-investor-day, accessed February 10, 2015.

36. Tiqqun, *This Is Not a Program*, trans. Joshua David Jordan (Cambridge, MA: MIT Press, 2011), 12.

Chapter 4

1. Deleuze, "Postscript on Control Societies," 179.

2. For a discussion of files as media technologies of law, see Cornelia Vismann, *Files: Law and Media Technology*, trans. Geoffrey Winthrop-Young (Stanford, CA.: Stanford University Press, 2008). For a study of the card index as an example of data-processing ontology before the electronic digital computer, see Markus Krajewski, *Paper Machines: About Cards and Catalogs, 1548–1929* (Cambridge, MA: MIT Press, 2011).

3. See Sybille Krämer, "Writing, Notational Iconicity, Calculus: On Writing as a Cultural Technique," trans. Anita McHesney, *MLN* 118.3 (2003): 518.

4. See Franz Kafka, *Letters to Felice*, ed. Erich Heller and Jürgen Born, trans. James Stern and Elizabeth Duckworth (New York: Schocken Books, 1973), 167–168, and *Letters to Milena*, ed. Willy Haas, trans. Tania Stern and James Stern (New York: Schocken Books, 1953), 229. For discussion of the first of these letters, see Gilles Deleuze and Felix Guattari, *Kafka: Toward a Minor Literature*, trans. Brian Massumi (Minneapolis: University of Minnesota Press, 1986), 94 n. 5; Kittler, *Gramophone, Film, Typewriter*, 57; and Bernhard Siegert, *Relays: Literature as an Epoch of the Postal Service*, trans. Kevin Repp (Stanford, CA: Stanford University Press, 1999), 256. For discussion of the second letter, see Deleuze and Guattari, *Kafka*, 30, and Kittler, *Gramophone, Film, Typewriter*, 225–226.

5. Deleuze and Guattari, *Kafka*, 81.

6. Kittler, *Gramophone, Film, Typewriter*, 3.

7. Ibid., 1.

8. Friedrich Kittler, *Discourse Networks 1800/1900*, trans. Michael Metteer, with Chris Cullings (Stanford, CA: Stanford University Press, 1990), 370.

9. Beller, *Cinematic Mode of Production*, 150–155.

10. Kittler, *Discourse Networks 1800/1900*, 370.

11. It should be recalled that Jacques Derrida and Catherine Malabou, among others, include the symbolic systems of structuralist linguistics, cybernetics, and genetics under the rubric "writing." See Jacques Derrida, *Of Grammatology*, trans. Gayatri Chakravory Spivak (Baltimore: Johns Hopkins University Press, 1976), 9, and Catherine Malabou, *Plasticity at the Dusk of Writing: Dialectic, Destruction, Deconstruction*, trans. Carolyn Shread (New York: Columbia University Press, 2010), 57. In Kittler's project of mapping media forms onto Jacques Lacan's real-imaginary-symbolic triad, literature belongs with the computer in the realm of the symbolic (rather than with analog audio recording and the real or with film and the imaginary). Kittler's Lacanian media triad is discussed at greater length in chapter 5.

12. Agamben, *What Is an Apparatus?*, 14.

13. Claude Shannon and Warren Weaver, *The Mathematical Theory of Communication* (Urbana: University of Illinois Press, 1949), 3.

14. See Heinz von Foerster, Margaret Mead, and Hans Lukas Teuber, eds., *Cybernetics: Transactions of the Eighth Conference, March 15–16, 1951* (New York: Josiah Macy Jr. Foundation, 1952), 22. For a discussion of this specific exchange between Shannon and Bavelas, see Heims, *The Cybernetics Group*, 221, and Hayles, *How We Became Posthuman*, 55.

15. "The Basic English vocabulary is limited to 850 words and the redundancy is very high. This is reflected in the expansion that occurs when a passage is translated into British English. Joyce on the other hand enlarges the vocabulary and is alleged to achieve a compression of semantic content" (Shannon and Weaver, *The Mathematical Theory of Communication*, 56). For a lengthier discussion of Shannon, Ogden, and Joyce, see Lydia Liu, "iSpace: Printed English after Joyce, Shannon, and Derrida," *Critical Inquiry* 32.3 (2006): 516–550.

16. Von Neumann and Morgenstern, *Theory of Games and Economic Behavior*, 9, 15, 31, 87, 555.

17. Ibid., 176–177.

18. Wiener, *The Human Use of Human Beings*, 84, 96–97, 114.

19. See Chun, *Programmed Visions*, 161. Chun cites Poundstone, *Prisoner's Dilemma*, (New York: Doubleday, 1992), 194.

20. Chun, *Programmed Visions*, 161. See also Nicholas Vonneumann, "The Philosophical Legacy of John von Neumann, in the Light of Its Inception and Evolution in His Formative Years," in James Glimm, John Impagliazzo, and Isadore Manuel Singer, eds., *The Legacy of John von Neumann* (Providence, RI: American Mathematical Society, 1990), 20. Although this name appears as "Vonneuman" in the cited text, I have changed it to "von Neumann" in my analysis in order to retain consistency across the surnames of the two brothers.

21. Chun, *Programmed Visions*, 162. The edition of *Faust* Chun cites is the one translated by David Luke (Oxford: Oxford University Press, 1987), ll. 1225–1237.

22. Norbert Wiener and Karl Deutsch, "The Lonely Nationalism of Rudyard Kipling," *Yale Review* 52.4 (1963): 502.

23. Elliot L. Gilbert, "The Lonely Nationalism of Rudyard Kipling" (review article), *Kipling Journal* 46.212 (1979): 13.

24. For the version of this essay presented at the eighth Macy conference, see I. A. Richards, "Communication between Men: Meaning of Language," in von Foerster, Mead, and Teuber, eds., *Cybernetics: Transactions of the Eighth Conference*, 45–91. The paper was later published as "Towards a More Synoptic View" in I. A. Richards, *Speculative Instruments* (London: Routledge and Kegan Paul, 1955), 113–126, which is quoted from here.

25. Richards, "Towards a More Synoptic View," 113, 123–124.

26. Margaret Morse, "An Ontology of Everyday Distraction: The Freeway, the Mall, and Television," in Patricia Mellencamp, ed., *Logics of Television: Essays in Cultural Criticism* (Bloomington: Indiana University Press, 1990), 190.

27. Krämer, "Writing, Notational Iconicity, Calculus," 519; Kittler, *Gramophone, Film, Typewriter*, 1–19.

28. K-Punk [Mark Fisher], "Cartesianism, Continuum, Catatonia: Beckett," http://k-punk.abstractdynamics.org/archives/007587.html, accessed 2/7/2014, quoted in McKenzie Wark, *Gamer Theory* (Cambridge, MA: Harvard University Press, 2007), paragraph 223, with "cyberspace" changed to "gamespace" throughout.

29. See the texts on cognitive capitalism and immaterial labor discussed in the introduction and chapter 1, as well as Christian Marazzi, *Capital and Affects: The Politics of the Language Economy*, trans. Guiseppina Mecchia (Cambridge, MA: MIT Press, 2011).

30. Mel Gussow, "Beckett at 75: An Appraisal," *New York Times*, April 19, 1981.

31. Jacques Derrida, "Two Words for Joyce," in Derek Attridge and Daniel Ferrer, eds., *Post-structuralist Joyce: Essays from the French* (Cambridge: Cambridge University Press, 1984), 147–148.

32. Mark Nixon writes that Beckett read Leibniz's *Monadology* in December 1933 and in the same month wrote in a letter to Tomas MacGreevy that he found the philosopher "a great cod, but full of splendid little pictures" (Mark Nixon, *Beckett's German Diaries, 1936–1937* [London: Continuum, 2011], 162). Also see Chris Ackerley, "Monadology: Samuel Beckett and Gottfried Wilhelm Leibniz," in Matthew Feldman and Karim Mamdani, eds., *Beckett/Philosophy* (Sofia, Bulgaria: University Press St. Klimrnt Ohridski, 2012), 122–145. Wiener's enthusiasm for Leibniz is well documented, not least in his own book *Cybernetics*, where he writes:

If I were to choose a patron saint of cybernetics out of the history of science, I should have to choose Leibniz. The philosophy of Leibniz centers about two closely related concepts—that of a universal symbolism and that if a calculus of reasoning. Now, just as the calculus of arithmetic lends itself to a mechanization progressing through the abacus and the desk computing machine to ultra-rapid computing machines of the present day, so the *calculus ratiocinator* of Leibniz contains the germs of *machina ratiocinatrix*, the reasoning machine. Indeed, Leibniz himself, like his predecessor Pascal, was interested in the construction of computing machines in the metal. It is therefore not in the least surprising that the same intellectual impulse which has led to the development of mathematical logic has at the same time led to the ideal or actual mechanization of thought. (12)

33. Marx, *Economic and Philosophic Manuscripts of 1844*, 302.

34. Manovich, *The Language of New Media*, 180.

35. Gilles Deleuze, "The Exhausted," in *Essays Critical and Clinical*, trans. Daniel W. Smith and Michael A. Greco (London: Verso, 1998), 156–159.

36. Alain Badiou also identifies a movement toward images in Beckett following *How It Is*. He finds the preceding forms "progressively replaced" with what he deems *"the figural poem of the subject's postures."* Badiou notes that this replacement is a definite progression from the previous works, which are continuous with Kafka's writing, supporting the periodization that places Kafka at the transition between disciplinary societies and control societies and Beckett alongside the ongoing development of control (*On Beckett*, ed. Alberto Toscano and Nina Power [Manchester, UK: Clinamen, 2003], 16).

37. "Watt began, and ended, in Paris: the first entries in what would prove to be six notebooks dated 'II February 1941,' and the last signed off with 'Dec 28th 1944 / End.' Much of the writing was done while Beckett was on the run from the Gestapo between 1943 and 1945, in the small town of Roussillon, in the Vaucluse, where he and his partner, Suzanne, had taken refuge" (Chris Ackerley, "Preface," in Samuel Beckett, *Watt* [London: Faber & Faber, 2009], vii).

38. Hugh Kenner, *The Mechanic Muse* (New York: Oxford University Press, 1987), 91–92.

39. Pascal, an imperative, procedural programming language developed by Niklaus Wirth in 1968, which forms the basis of early Apple Macintosh assembly languages, is based on the ALGOL (ALGOrithmic Language) family, which is especially suited to pseudocode examples for the written description of algorithms. See Niklaus Wirth, "The Programming Language PASCAL," *Acta Informatica* 1 (1971): 35–63, and the materials collected at Scott Moore's ISO 7185 Standard Pascal Page, http://www .standardpascal.org, accessed February 7, 2014.

40. Beckett, *Watt*, 119–120. This passage is quoted without publication information in Kenner, *The Mechanic Muse*, 93, so references are to the 2009 Faber & Faber edition cited in note 37.

41. Kenner, *The Mechanic Muse*, 94.

42. Ibid., 95. Kenner was an able computer programmer as well as a literary critic, contributing a column to the computing magazine *Byte* in the early 1980s. See Harvey Blume, "Hugh Kenner: The Grand Tour," http://www.harveyblume. com/1998/04/q-hugh-kenner-grand-tour.html, accessed February 7, 2014.

43. S. E. Gontarski and Chris Ackerley, "Samuel Beckett's *Watt*," in Brian W. Shaffer, ed., *A Companion to the British and Irish Novel 1945–2000* (Oxford: Wiley, Blackwell, 2005), 233. See also Hugh Kulik, "Mathematics as Metaphor: Samuel Beckett and the Esthetics of Incompleteness," *Papers on Language and Literature* 29 (1993): 131–151, and Howard J. Alane, "The Roots of Beckett's Aesthetic: Mathematical Allusions in *Watt*," *Papers on Language and Literature* 30 (1994): 346–351.

44. Beckett, *Watt*, 72–84.

45. Claude Shannon, "Prediction and Entropy of Printed English," *Bell Systems Technical Journal* 30.1 (1951), 58.

46. See Friedrich Kittler, "Code," in Matthew Fuller, ed., *Software Studies: A Lexicon* (Cambridge, MA: MIT Press, 2008), 40–47, for a discussion of computer code in relation to the history of cryptography.

47. Kittler, "Real Time Analysis, Time Axis Manipulation," 192.

48. See Rubin Rabinovitz, *The Development of Samuel Beckett's Fiction* (Urbana: University of Illinois Press, 1984), 153, and Barbara Reich Gluck, *Beckett and Joyce* (London: Bucknell University Press, 1979), 91–92. Turing's article "On Computable Numbers" sets out the definitions of countable and hence computable numbers for a theoretical computing machine, although in "There Is No Software" Kittler queries the consistency of Turing's definitions (189 n. 13).

49. "Having ascertained these pivotal yesses and noes, the program, like the sentence, dies away a little feebly, since it tells us nothing about what happened when Mrs. Gorman came, or about her sensations when she did not come. Is she snug, in her chair, by the fire? Elated, in her chair, by the open window?" (Kenner, *The Mechanic Muse*, 95–96).

50. Ibid., 95.

51. Samuel Beckett, *Molloy, Malone Dies, the Unnamable* (London: Calder, 1994), 92.

52. Ibid., 98.

53. Ibid., 103, emphasis added.

54. Gilles Deleuze, *Difference and Repetition*, trans. Paul Patten (New York: Columbia University Press, 1994), 79.

55. Galloway has made the same point in responding to Chun and Kittler, stating that "it is foolish to think that writing an 'if/then' control structure in eight lines of assembly code is any more or less machinic than doing it in one line of C, just as the same quadratic equation may swell with any number of multipliers and still remain balanced. The relationship between the two is *technical*" ("Language Wants to Be Overlooked," 319).

56. See Kittler, "There Is No Software," and Chun, "On Software," 26–51.

57. Herman H. Goldstine and John von Neumann, *Planning and Coding of Problems for an Electronic Computing Instrument* (Princeton, NJ: Institute for Advanced Study, 1948).

58. The play's history is given in S. E. Gontarski, ed., *The Theatrical Notebooks of Samuel Beckett*, vol. 2: *Endgame* (London: Faber and Faber, 1992), xvi.

59. Samuel Beckett, *Endgame*, in *Complete Dramatic Works* (London: Faber & Faber, 1990), 105, quoted in Kenner, *The Mechanic Muse* 100–101.

60. Kenner, *The Mechanic Muse*, 100.

61. Friedrich Kittler, "Protected Mode," in *Literature, Media, Information Systems*, 158. The publication Kittler refers to is B. Edlinger, H. G. Eichholtz, H. Feichtinger, J. P. Jordan, and U. Kern, *Chip-Tool-Praxis: Assembler-Programming Auf Dem PC*, issue 1 (1989), 9.

62. The crucial paper outlining the FFT's practical applications is J. W. Cooley and J. W. Tukey, "An Algorithm for Machine Calculation of Complex Fourier Series," *Mathematical Computation* 19.90 (1965): 297–301. For a commanding history of techniques for the digitization of audio see Jonathan Sterne, *MP3: The Meaning of a Format* (Durham, NC: Duke University Press, 2012).

63. See Ivan Edward Sutherland, "Sketchpad, a Man–Machine Graphical Communication System," electronic version available at http://www.cl.cam.ac.uk/techreports/UCAM-CL-TR-574.pdf, accessed February 7, 2014.

64. Beckett, *Quad*, in *Complete Dramatic Works*, 451.

65. Ibid.

66. Elizabeth Klaver has made a similar remark on *Quad*, stating that it presents a "cycle of repetition which is similar to a computer program in loop mode" ("Samuel Beckett's *Ohio Impromptu*, *Quad*, and *What Where*: How It Is in the Matrix of Text and Television," *Contemporary Literature* 32.3 [1991]: 376).

67. Frank Gray, "Pulse Code Communication," U.S. Patent 2,632,058, filed November 13, 1947, awarded March 17, 1953.

68. See Frank Gray, J. W. Horton, and R. C. Mathes, "The Production and Utilization of Television Signals," *Transactions of the AIEE* 46 (June 1927): 560–603; Frank

Gray, "The Use of a Moving Beam of Light to Scan a Scene for Television," *Journal of the Optical Society of America* 16.3 (1928): 177–185; Herbert E. Ives, Frank Gray, and M. W. Baldwin, "Image Transmission System for Two-Way Television," *Bell System Technical Journal* 9.3 (July 1930): 448–469; and Pierre Mertz and Frank Gray, "A Theory of Scanning and Its Relation to the Characteristics of the Transmitted Signal in Telephotography and Television," *Bell System Technical Journal* 13.3 (1934): 464–515.

69. Beckett, *Quad*, 453.

70. Deleuze, *Essays Critical and Clinical*, 158.

71. Ibid., 154.

72. For more on the informatic capture of human actions, see Agre, "Surveillance and Capture."

73. Tiqqun, "The Cybernetic Hypothesis," 47–48, emphasis in original.

Chapter 5

1. Dupuy, *On the Origins of Cognitive Science*, 156; Agamben, *What Is an Apparatus?* 21.

2. Jameson, *The Political Unconscious*, 64.

3. Warren Sack, "Memory," in Fuller, ed., *Software Studies*, 189, emphasis added.

4. B. N. Colby, George A. Collier, and Susan K. Postal, "Comparison of Themes in Folktales by the General Inquirer System," *Journal of American Folklore* 76.302 (1963): 318.

5. It is worth reiterating here that, for Tiqqun, the political foundation of the cybernetic hypothesis guiding late capitalism is its conceptualization of "biological, physical, and social behaviors as … integrally programmed and reprogrammable"— a procedure that requires a disqualification of "all 19th century psychology, including psychoanalysis," as a "myth" ("The Cybernetic Hypothesis," 42, 49).

6. Wendy Hui Kyong Chun, "Programmability," in Fuller, ed., *Software Studies*, 225.

7. Ibid.

8. Ibid.

9. Foucault, *The Birth of Biopolitics*, 223.

10. Manuel Castells, "Materials for an Exploratory Theory of the Network Society," *British Journal of Sociology* 51.1 (2001): 12.

11. The literature on NLP is vast. For examples of such guidebooks, see Lynne Cooper, *Business NLP for Dummies* (Chichester, UK: Wiley, 2008); Pat Hutchinson, *How to Sell with NLP* (London: Pearson Business, 2010); Joseph O'Connor, *NLP and Sports: How to Achieve Your Own Peak Performance* (New York: Harper Thorsons, 2001); Ross Jeffries, *How to Get the Women You Desire into Bed* (n.p.: self-published, 1992); and Neil Strauss, *The Game: Penetrating the Secret Society of Pickup Artists* (New York: Regan Books, 2005).

12. For the methodological foundation texts for the development of NLP, see Richard Bandler and John Grinder, *The Structure of Magic I: A Book about Language and Therapy* (Palo Alto, CA: Science and Behavior Books, 1975); *The Structure of Magic II: A Book about Communication and Change* (Palo Alto, CA: Science and Behavior Books, 1975); *Patterns of the Hypnotic Techniques of Milton H. Erickson, M.D.*, vol. 1 (Cupertino, CA: Meta, 1976); and, with Judith Deloizer, *Patterns of the Hypnotic Techniques of Milton H. Erickson, M.D.*, vol. 2 (Cupertino, CA: Meta, 1977). The first book-length presentation of NLP under that name is John Grinder and Richard Bandler, *Frogs into Princes: Neuro Linguistic Programming* (Layafette, CA: Real People Press, 1979).

13. Bandler and Grinder, *The Structure of Magic I*, 1–2.

14. In a 2001 text, Carmen Bostic St. Clair and John Grinder describe the practice as "a modeling technology whose specific subject matter is the set of differences that make the difference between the performance of geniuses and that of average performers in the same field or activity. In this sense, the objective of modeling studies in NLP is to explicate in a transferable and learnable code these sets of differences" (*Whispering in the Wind* [Palo Alto, CA: J&C Enterprises, 2001], n.p.).

15. For an account of Erickson's and Bateson's respective involvement with the founders of NLP, see Paul Tosey and Jane Mathison, *Neuro-Linguistic Programming: A Guide for Managers and Developers* (Basingstoke, UK: Palgrave Macmillan, 2009), 70–84, 85–96. Bateson's relationship to the NLP founders appears to be what Tiqqun refer to when they state that he was involved in the origins of the "'sales techniques training' movement developed at Palo Alto" ("The Cybernetic Hypothesis," 49).

16. Gregory Bateson, introduction to Bandler and Grinder, *The Structure of Magic I*, x.

17. Lévi-Strauss, "Language and the Analysis of Social Laws," 61.

18. Friedrich Kittler, "The World of the Symbolic, a World of the Machine," trans. Stefanie Harris, in *Literature, Media, Information Systems*, 135, 141.

19. Ibid., 141.

20. This line of inquiry is not without precedent. Lydia H. Liu, for example, suggests that Lacan's main contribution "lies in what he can tell us about the cybernetic unconscious of the postwar Euro-American world order" ("The Cybernetic

Unconscious: Rethinking Lacan, Poe, and French Theory," *Critical Inquiry* 36 [Winter 2010]: 289).

21. See Jodi Dean, *Blog Theory: Feedback and Capture in the Circuits of Drive* (Cambridge: Polity, 2010), and Galloway, *The Interface Effect*, 27–28.

22. See, for example, Dupuy, *On the Origins of Cognitive Science*, 22, 79–80, 85–86, 91, 108–109; Hayles, *How We Became Posthuman*, 70–73; Celine Lafontaine, *L'Empire cybernétique: Des machines à penser à la pensée machine* (Paris: Seuil, 2004); part 1 of John Johnston, *Allure of Machinic Life: Cybernetics, Artificial Life, and the New AI* (London: MIT Press, 2008); Liu, "The Cybernetic Unconscious" and the book in which it is collected, *The Freudian Robot: Digital Media and the Future of the Unconscious* (Chicago: University of Chicago Press, 2011).

23. Deutsch, *The Nerves of Government*, 79.

24. Dupuy, *On the Origins of Cognitive Science*, 18–19.

25. See Heims, *The Cybernetics Group*, 135–139, and Hayles, *How We Became Posthuman*, 71–73.

26. Hayles, *How We Became Posthuman*, 70–73.

27. For Kubie's work on reverberating circuits, see Lawrence Kubie, "A Theoretical Application to Some Neurological Problems of the Properties of Excitation Waves Which Move in Closed Circuits," *Brain* 53 (1930): 166–177, and "The Repetitive Core of Neurosis," *Psychoanalytic Quarterly* 10 (1941): 23–43. McCulloch recounts the importance of the first of these papers in "Recollections," 12, and this influence is reiterated in Hayles, *How We Became Posthuman*, 70.

28. Heims, *The Cybernetics Group*, 126–127.

29. Kubie, "Fallacious Use of Quantitative Concepts," 517.

30. Sigmund Freud, *Introductory Lectures on Psychoanalysis* (London: Allan & Unwin, 1922), 313, quoted in ibid., 507–508.

31. Kubie, "Fallacious Use of Quantitative Concepts," 508.

32. Ibid., 516.

33. Ibid., 517–518.

34. Dupuy describes Kubie as "indefatigable" in his battle for the retention of the unconscious within cybernetics debates and gives a statement by Rosenblueth as typical of others' response: "Either a mental event occurs or it does not. Either a train of nerve impulses is registered or it is not. To say that one has been registered only 'unconsciously' is nonsense" (quoted in *On the Origins of Cognitive Science*, 83).

35. Heims, *The Cybernetics Group*, 126.

36. Dupuy writes that Lacan's knowledge of cybernetics was "quite detailed" in some areas, noting that he "took an interest … in the theory of closed reverberating circuits that Lawrence Kubie's work in the 1930s had led McCulloch to take up, and he was familiar with the work of the British neuroanatomist John Z. Young, discussed at the ninth Macy Conference in March 1952, testing this theory in the octopus" (*On the Origins of Cognitive Science*, 109). Liu also notes Lacan's awareness of Young's work on the neural networks of the octopus in *The Freudian Robot*, 136.

37. Jacques Lacan, *The Seminar of Jacques Lacan, Book II: The Ego in Freud's Theory and in the Technique of Psychoanalysis 1954–1955*, trans. Sylvana Tomaselli (New York: Norton, 1991), 179. In "The Cybernetic Unconscious" (299–311), Liu traces this reference to Poe to the work of Georges Théodule Guilbaud, specifically the review essay "La Théorie des jeux: Contributions critiques a` la théorie de la valeur," *Économie Appliquée* 2 (1949): 209–306. Guilbaud's paper is translated by A. L. Minkes as "The Theory of Games: Critical Contributions to the Theory of Value," in Mary Ann Dimand and Robert W. Dimand, eds., *The Foundations of Game Theory* (Cheltenham, UK: Edward Elgar, 1997), 348–376. Geoghegan argues that both Johnston's and Liu's accounts are overly concerned with written accounts of cybernetics and thus neglect material technologies such as the automata produced by Shannon and David Hagelberger as well as the specific historical and institutional arrangements that surrounded the meeting of cybernetics and "French theory" ("From Information Theory to French Theory," 100 n.11).

38. Lacan, *The Seminar of Jacques Lacan, Book II*, 120.

39. Ibid.

40. Ibid., 192.

41. Ibid., 193.

42. Ibid., 192–193, emphasis added.

43. Kittler is here referring to the statistical forecasting of Markov chains, "memoryless" series of states of which the McCulloch–Pitts neuron, cellular automata, and dice rolls are examples. "The input to the symbolic machine is a throw of the dice in the real because the French *dé* [dice], to the delight of Mallarmé and Lacan, is derived from the Latin *datum*" ("World of the Symbolic," 141).

44. Ibid.

45. Jon Ronson, *The Men Who Stare at Goats* (New York: Simon & Schuster, 2004).

46. Kittler, "World of the Symbolic," 130.

47. Ibid.

48. Ibid.

49. See, for example, Brenda Laurel, *Computers as Theater* (Reading, MA: Addison-Wesley, 1991), 51, and Michael Mateas, "A Preliminary Poetics for Interactive Drama and Games," in Noah Wardrip-Fruin and Pat Harrigan, eds., *First Person: New Media as Story, Performance, and Game* (Cambridge, MA: MIT Press, 2004), 22, 24.

50. This lack of resolution in *The Parallax View*, centered on the partial, fragmented representation of the conspiracies, is addressed in Mark Fisher, *Capitalist Realism: Is There No Alternative?* (Ropley, UK: 0 Books, 2009), 67.

51. Jameson, *The Geopolitical Aesthetic*, 16.

52. Ibid.

53. Ibid., 60.

54. Sybille Krämer, "The Cultural Techniques of Time Axis Manipulation: On Friedrich Kittler's Conception of Media," *Theory, Culture, and Society* 23.7–8 (2006): 94.

55. "The imaginary implements precisely those optical illusions that were being researched in the early days of cinema. A dismembered or (in the case of film) cut-up body is faced with the illusionary continuity of movements in the mirror or on screen. It is no coincidence that Lacan recorded infants' jubilant reactions to their mirror images in the form of documentary footage" (Kittler, *Gramophone, Film, Typewriter*, 15).

56. Friedrich Kittler, *Optical Media*, trans. Anthony Enns (Cambridge, UK: Polity, 2010), 162. Kittler puts this definition in broader media-historical terms later in the same book: "To begin with, computer technology simply means being serious about the digital principle. What are only the edits between frames in film...become the be all and end all of digital signal processing" (*Optical Media*, 225). In this regard, Kittler is at least partly in agreement with Lev Manovich, who argues that cinema is fundamentally digital, being composed of discrete still images that are sampled from a continuous flow of time. See Manovich, *Language of New Media*, 50–51.

57. Galloway, *Gaming*, 94.

58. David Bordwell, "A Glance at Blows," http://www.davidbordwell.net/blog/2008/12/28/a-glance-at-blows/, accessed February 7, 2014), and Steven Shaviro, *Post-cinematic Affect*, 187 n. 83. Given the laborious nature of film editing, it is easy to favor Shaviro's reading over Bordwell's because Shaviro stresses a reading of form as purposefully created and composed rather than thrown together. The latter seems an unlikely procedure given the precise economic logic that underpins all aspects of production on such a project.

59. Sean Cubitt, *The Cinema Effect* (Cambridge, MA: MIT Press, 2004), 230.

60. Bateson, *Steps to an Ecology of Mind*, 271. It should come as no great surprise that a manual on the advantages to be gained from using NLP in the workplace is named

for Bateson's claim. See Sue Knight, *NLP at Work: The Difference That Makes the Difference in Business* (Boston, MA: Nicholas Brealey, 2002).

61. Jean-François Lyotard, "Acinema," in Philip Rosen, ed., *Narrative, Apparatus, Ideology: A Film Theory Reader* (New York: Columbia University Press, 1986), 350, 352.

62. Mark Fisher has made a similar observation, noting that the *Bourne* films construct a temporality of the "continuous present," a "series of evanescent event-ciphers and action set pieces which barely cohere into an intelligible narrative" (*Capitalist Realism*, 58).

63. Gilles Deleuze, *Cinema 2*, trans. Hugh Tomlinson and Robert Galeta (London: Athlone, 1989), 213–214.

64. One recalls here both Rosenblueth, Wiener, and Bigelow's account of phenomena "so fast that it is not likely that nerve impulses would have time to arise at the retina, travel to the central nervous system and set up further impulses which would reach the muscles in time to modify the movement effectively" ("Behavior, Purpose, and Teleology," 20) and Ronald Reagan's pronouncement in an August 8, 1983 speech that "many young people have developed incredible hand, eye, and brain coordination" in playing video games and that "the air force believes these kids will be our outstanding pilots should they fly our jets."

65. Friedrich Kittler, "Computer Graphics: A Semi-technical Introduction," *Grey Room* 2 (Winter 2001): 31.

66. Wolfgang Ernst, "Else Loop Forever: The Untimeliness of Media," http://www.medienwissenschaft.hu-berlin.de/medientheorien/downloads/publikationen/ernst-else-loop-forever.pdf, accessed February 7, 2014.

67. Wark, *Gamer Theory,* paragraph 091.

68. Kittler, *Gramophone, Film Typewriter*, 259.

69. "In its nature, the door belongs to the symbolic order, and it opens up either on to the real, or the imaginary, we don't know quite which, but it is either one or the other. There is an asymmetry between the opening and the closing—if the opening of the door controls access, when closed, it closes the circuit" (Lacan, *The Seminar of Jacques Lacan, Book II*, 302). For a longer, historically nuanced discussion of Lacan's theorization of the door, see Bernhard Siegert, "Doors: On the Materiality of the Symbolic," trans. John Durham Peters, *Grey Room* 47 (Spring 2012): 6–23.

70. For a longer account of the differences between targeting and striking in *Counter-Strike* (a multiplayer first-person shooter based on the *Half-Life* engine) and *World of Warcraft*, see Alexander R. Galloway, "Starcraft, or, Balance," *Grey Room* 28 (Summer 2007): 88–91.

71. This claim is, of course, resonant with Rey Chow's concept of the world target. One of the present book's aims is to provide a conceptual, technological, and aesthetic genealogy for this rich historical concept. See Chow, *Age of the World Target* (Durham, NC: Duke University Press, 2006).

72. Jameson, *Political Unconscious*, 6.

73. In *Anti-Oedipus*, Deleuze and Guattari write that "the [critical] opposition is between the class and those who are outside the class" (*Anti-Oedipus*, trans. Robert Hurley, Mark Seem, and Helen R. Lane [Minneapolis: University of Minnesota Press, 1983], 255). For a lengthier engagement with this aspect of Deleuze and Guattari's work, see Fredric Jameson, "Marxism and Dualism in Deleuze," *South Atlantic Quarterly* 96.3 (1997): 393–416.

74. Deleuze, "Control and Becoming," 175, and "Postscript on Control Societies," 180.

75. Deleuze, "Control and Becoming," 176.

76. Giorgio Agamben, *The Coming Community*, trans. Michael Hardt (1990; translation, Minneapolis: University of Minnesota Press, 2007), 85, 5, 85. For more on this concept of unrepresentability, see Augusto Illuminati, "Unrepresentable Citizenship," in Hardt and Virno, eds., *Radical Thought in Italy*, 167–188.

77. Tiqqun, "The Cybernetic Hypothesis," 72, 73.

78. Alexander R. Galloway and Eugene Thacker, *The Exploit*, 174.

79. Wendy Hui Kyong Chun, "Crisis, Crisis, Crisis, or Sovereignty and Networks," *Theory, Culture and Society* 28.6 (2011): 92.

Index

Printed in the United States
by Baker & Taylor Publisher Services